"To Subsidise My Income"

Afrika-Studiecentrum Series

VOLUME 7

"To Subsidise My Income"

Urban Farming in an East-African Town

By

Dick Foeken

BRILL
LEIDEN · BOSTON
2006

Cover photo: Sam Owuor

This book is printed on acid-free paper.

Library of Congress Cataloging-in-Publication Data

Foeken, D.
To subsidise my income : urban farming in an East-African town / by Dick Foeken in association with Samuel O. Owuor ... [et al.].
p. cm. — (Afrika-Studiecentrum series ; 7)
Includes bibliographical references (p.).
ISBN-13: 978-90-04-15202-1
ISBN-10: 90-04-15202-4 (pbk. : alk. paper)
1. Urban agriculture—Kenya—Nakuru. I. Title. II. Series.

S473.K4F64 2006
630.9173'2—dc22

2006043929

ISSN 1570–9310
ISBN-13: 978-90-04-15202-1
ISBN-10: 90-04-15202-4

PRINTED IN THE NETHERLANDS

This book is dedicated to

Marijke

Contents

Maps

Figures

Boxes

Tables

Photos

Glossary and abbreviations

4-K (Club)	*Kuungana, Kufanya, Kusaidia, Kenya* ("Get together, act and help Kenya")
ARDP	Agricultural and Rural Development Programme
CBD	central business district
CBO	community-based organisation
CBS	Central Bureau of Statistics (Nairobi)
Cd	cadmium
CDN	Catholic Diocese of Nakuru
chang'aa	local brew, usually made of maize
debe	measurement comparable with a bucket (17-18 kg)
dhania	parsley
ECLOF	Ecumenical Church Loan Fund
g	gram(s)
ha	hectare(s)
HAZ	height-for-age, median z-score
Hg	mercury
jembe	kind of hoe
kcal	kilocalories
kg	kilogramme(s)
kienyeji	local chicken
Ksh.	Kenyan shilling(s)
kunde	peas
l	litre(s)
LA21	Localising Agenda 21
m^2	square metre(s)
maharague	beans
mahindi	maize
managu	black night shade
matatu	type of joint taxi
mavaki	local vegetable
mbiriganya	egg plant
mchicha	'wild spinach'
MCN	Municipal Council of Nakuru
mg	milligram(s)
NGO	non-governmental organisation

NSFCK	National School Feeding Council of Kenya
NUAP	Nakuru Urban Agriculture research Project
panga	kind of cutlass
Pb	lead
ppm	parts per million
RVIST	Rift Valley Institute of Science and Technology
saget	spider plant
SAMCAF	Southern Africa Microfinance Capacity Building Facility
sh.	(Kenyan) shilling(s)
shamba	plot
sukuma wiki	kale
terere	amaranthus
TOL	temporary occupation license
ugali	stiff maize porridge
UNCHS (Habitat)	United Nations Centre for Human Settlements
US$	US dollar(s)
WAZ	weight-for-age, median z-score
WHZ	weight-for-height, median z-score
WHO	World Health Organisation
WMS III	Welfare Monitoring Survey 1997
Zn	zinc

Foreword

> *John and Mary arrived in Nakuru in 1971 and settled in Lakeview where they still live today. They have always cultivated maize and beans for home consumption in town on a plot measuring 100x50 feet. In 1999, they grew kale and arrowroot as well. They used no chemicals and did not irrigate the crops. They only harvested in 1999 a small amount because of the drought. Mary dug up some arrowroot each month. She did not sell anything but gave away a few cobs of maize to her children. Mary is the one responsible for farming and does it all on her own, spending on average two hours a day on the* shamba[1] *all year round. The main impact of the drought was that she had to spend more money on food than in normal years. Cultivating crops in town is important for her because "it helps to feed my family". It has become more important for her over the years because "now food is expensive but I'm able to deal for that with what I harvest". She gets something from the* shamba *throughout the whole year. So she would never stop with her farming activities because it helps "to subsidise my income".*

This describes in a nutshell the importance of urban farming for a low-income household like that of John and Mary. Even though plots are usually small and production is quite modest, it does help these people in terms of food, and indirectly also income. This book is about such people.

Farming in town, however, is not without its critics. For many urban managers and policy-makers, agriculture is simply not an acceptable type of urban land use. On official maps showing urban land uses, agriculture is seldom included. Moreover, urban managers are inclined to stress its negative environmental aspects, such as pollution, erosion, noise, smell, etc. They see urban agriculture as a menace that should preferably be forbidden. Others, however, including many researchers, stress the importance of the activity for the livelihood of urban residents (the poor in particular), as well as the (potentially) positive contribution urban agriculture can make towards urban food supply, the

1 *Shamba* is the Swahili word for plot or piece of land.

management of the urban environment – for example in terms of waste management and recycling – and its aesthetic attraction.

Urban farming is not only an important but also a complex phenomenon, with economic, social, cultural, political, environmental as well as legal aspects. This book discusses urban farming from this broad perspective, even though it is not possible to treat all aspects at length. The studies on which it is based took place in Nakuru, Kenya, a medium-sized town, of which there are so many in sub-Saharan Africa. A whole range of questions are answered in this book: Who are the farmers of Nakuru town? What crops do they grow and how? What types of livestock are kept and how? What are the benefits for Nakuru as a whole and for the people involved? To what extent are the farmers supported and to what effect? What is the environmental impact of farming in Nakuru? In addition, a specific form of institutional urban agriculture is also discussed, namely school farming. Finally, special attention is devoted to the poorer segments of Nakuru's population. Do they participate in urban farming and if so, to what extent do they benefit from it?

The research team received an enthusiastic welcome from the relevant local officials when this study on urban agriculture in Nakuru was originally proposed in 1998. In the context of the Localising Agenda 21 programme, urban planning was being developed, and systematic knowledge of urban agriculture was lacking. It was therefore decided to start the research project, which was being known as the Nakuru Urban Agriculture Research Project (NUAP), with a general survey in 1999 to obtain an overall picture of farming activities at town level. Besides this main study, a number of smaller studies were carried out in the following years, all of them dealing with a specific aspect of urban farming in Nakuru (see Annex 1).

In November 2002, a two-day workshop was organized in Nakuru to disseminate the results of all these studies among the local stakeholders and to formulate suggestions for improvements in the sector and recommendations for policy and planning. The present book reflects all the studies carried out in the context of NUAP, including the results of the workshop (see Annex 11).

Different researchers participated in the research project: Samuel O. Owuor, Wijnand Klaver, Correta E. Odera, Peter W. King'ori, Ernest O. Nyandwaro and Nicole Versleijen. Their efforts and enthusiasm are highly appreciated and are 'rewarded' by co-authorship of the chapter(s) concerned. Nicole Versleijen agreed to allow her thesis to be used throughout the book, for which I am very grateful. It not only enlivens the otherwise somewhat dry text but also shows that each urban farmer is actually an individual 'case'. A large number of local officials, representatives from local non-government organizations, urban farmers and assistants participated in some stage or in some specific study in the research project. They are too many to mention them individually by name, but all of them have been acknowledged in previous publications (Foeken & Owuor 2000; Foeken *et al.* 2002) and in forthcoming theses. One person has to be mentioned, however, namely my colleague, student and friend Samuel Ouma Owuor. Without his unconditional hard work and support at all stages of the project, NUAP would not have run as smoothly as it did.

Finally, I want to express my gratitude to Leo van den Berg for his detailed comments on the manuscript, to Nel de Vink for drawing the maps and to Ann Reeves for correcting what otherwise might have been my 'English translation of Dutch'.

1

The issue

Urbanization and urban poverty

In June 1996, the second world-wide Habitat Conference took place in Istanbul. The conference was entirely devoted to the rapidly increasing urban population in the world and the concomitant problems regarding urban management. In the beginning of the 20th century only 13% of the world's population were living in cities, but by the year 2010 over half of all people on earth will be urbanites (UNCHS 1996: 12). Especially in the Developing Countries, the urbanization process is taking place at an extremely fast rate. Although sub-Saharan Africa is still the least urbanized continent in the world, during recent decades it has known the most rapid growth of the urban population (UNCHS 1996: 84). It is expected that by the year 2020 about 60% of the population of sub-Saharan Africa will be living in urban areas.[1]

Besides natural growth, a major cause of the rapidly increasing urban population is the influx of migrants from the rural areas. Most of these migrants have only one way to go as soon as they have reached the city, namely to one of the slums or shantytowns where the urban poor live. Since the beginning of the 1980s in particular, these low income areas have grown substantially. It was estimated that in 1993 about 55% of the Nairobi population of about 1.5 million lived in these 'unplanned' and 'unserviced' areas (Gathuru 1993).

For many of these slum dwellers, it is very hard to find employment. Moreover, by the early 1990s, most African countries were implementing structural

---

[1] See http://www.fao.org/News/1997/970405-e.html.

adjustment programmes (SAPs), implying, amongst others, drastic cuts in public expenditures, trade liberalization, increased interest rates, etc. As a consequence, unemployment increased and real incomes fell, while at the same time prices rose and welfare services declined. Particularly the urban poor were hard hit (Tinker 1994; Drakakis-Smith *et al.* 1995).

Meanwhile, urban poverty is increasing dramatically in many African countries. Kenya is one of these countries. Most of these people live in slums or slum-like areas with limited access to basic services such as adequate water, schools and health services. Many of the urban poor have no regular work and, hence, no regular income. Moreover, a relatively large percentage of their income is spent on rent and food. As a result of their poverty, these people are excluded from credit facilities with which some kind of small business might be set up.

People's responses to (urban) poverty are twofold: first, they try to raise or at least maintain their income level and, secondly, they reduce their expenses. Raising or maintaining one's income can usually only be done by diversifying income sources, mainly in the informal sector. Livelihoods have become increasingly dependent on the informal sector and on casual work. Expenses may be cut in areas like education and health (all the more so because under structural adjustment these services have become virtually unaffordable for many of the poor), and cuts can be made on material expenses, as well as on consumption and dietary patterns.

Growing numbers of the urban poor engage in illicit income-generating activities. Hawking without a license and in forbidden areas is common. Some women engage in brewing prohibited liquor and in prostitution, in spite of the health risks involved. Drug dealing and peddling is on the increase as well (Kanji 1996). Another, often illegal activity that has become widespread is the production of food within the city or town limits. This is now an important coping mechanism in the context of cuts in food subsidies, increases in the cost of living and decreasing household purchasing power. Indeed, urban agriculture is by some seen as one of the means to realize Millennium Development Goal 1, which calls for a 50% reduction of poverty in the world by 2015 (see e.g. Mougeot 2005).

Urban agriculture in sub-Saharan Africa

Farming in town is a common feature of sub-Saharan Africa (Obudho & Foeken 1999). It was estimated that in the mid-1990s as much as 40% of the urban population in Africa was involved in urban agriculture (Mougeot 1994a).

Studies have been carried out across the continent[2] and from these, the following picture arises.[3]

Farming is undertaken wherever land is available. In built-up areas, this can be in one's own compound ('backyard farming' or 'on-plot farming') or on land belonging to someone else ('open space farming' or 'off-plot farming'), the owner being either the government or a private person. Farming is particularly common on the outskirts of urban centres, on formerly rural land that has now become part of the urban centre due to boundary extensions ('peri-urban farming'[4]). In these zones, both small-scale and large-scale farming can be found. However, as the urban centre grows, these areas gradually lose their rural character and farming becomes increasingly of the other two types.

Farming in town has increased enormously over the past two decades due to the economic crisis that prevailed in most African countries. For the poor, increasing their food security is usually the main motivation for farming in town, and for some it is even a survival strategy. Nevertheless, many of the poor also sell some of their produce, partly to be able to pay for other basic household needs, but also because some crops are perishable and cannot be stored

---

2 See e.g. on *Benin*: Brock 1999 (Cotonou); *Botswana*: Byerley 1996 (Gaborone); *Central African Republic*: Villien 1988 (Bangui); *Congo*: Vennetier 1961 (Pointe Noire); *Congo DR*: Mianda 1996 (Kinshasa); *Ethiopia*: Egziabher 1994 (Addis Ababa); *Ghana*: Armar-Klemesu & Maxwell 2000, Obosu-Mensah 1999 (Accra); *Guinea-Bissau*: Lourenço-Lindell 1996 (Bissau); *Mozambique*: Sheldon 1991; *Nigeria*: Tricaud 1987 and Gbadegesin 1991 (Ibadan), Ajaegbu *et al.* 2000 (Jos), Gefu 1992 (Zaria); *Senegal*: Mbaye & Moustier 2000 (Dakar); *Sierra Leone*: Tricaud 1987 (Freetown); *South Africa*: Baxter 1994, Rogerson 1994, Eberhard 1989 (Cape Town); *Tanzania*: Mosha 1991, Mlozi *et al.* 1992 (general); Dongus 2000, Jacobi *et al.* 2000, Mlozi 1996, Sawio 1993 and 1994 (Dar es Salaam); Foeken *et al.* 2004 (Mbeya, Morogoro); *Togo*: Schilter 1991 (Lomé); *Uganda*: Maxwell 1994 and 1995, Atakunda & Maxwell 1996 (all Kampala); *western Africa*: Diallo 1993; *Zambia*: Sanyal 1985, Rakodi 1988, Drescher 1996 (all Lusaka); *Zimbabwe*: Mbiba 1995, and Drakakis-Smith *et al.* 1995, ENDA-Zimbabwe 1996, Gumbo & Ndiripo 1996, Mbiba 2000 (all Harare).

3 Although many of the described characteristics of urban agriculture in sub-Saharan Africa apply to the whole region, the description fits better for the eastern and southern African regions than for the French-speaking western African region. In the latter area, urban farming tends to be somewhat less omnipresent, more of a commercial character and more of a men's business either or not organized as cooperatives.

4 'Peri-urban' is defined here as the zone between the built-up area and the town boundary. There exist other definitions in which the peri-urban zone extends further than the municipal boundary, hence including the rural areas around the urban centre. The usual criterion is whether production is for the urban market or not (see e.g. Jacobi 1997: 2). For simplicity and clarity reasons, we prefer the more limited definition of peri-urban.

and/or because storage space is unavailable. For medium-income and high-income households, commercial considerations are usually of more importance than among the poor, although the consumption of self-produced vegetables and milk is often highly valued. But for most of these households, the basic reason to do so is the same as for the poor, namely, as is often stated by the farmers themselves, "to subsidise my income".

The majority of African urban farmers are women. In most parts of Africa, women have traditionally been responsible for household food provision and farming is relatively easy to combine with the care of children. Women also often have lower educational levels than men, so it is difficult for them to compete in a shrinking labour market. Farming may, thus, be the only option left to them in a situation of unemployment and poverty. Several studies have found that the number of female-headed households is disproportionately high among urban farmers. It has also been shown that recent migrants often do not practise urban farming. A person has to be settled and have access to the right networks in order to be able to gain access to a plot of land.

The crops grown are mostly basic food crops such as maize, beans, cassava, sorghum, rice and yams. A wide range of vegetables is also cultivated, some of which are often sold because of their perishability and because there is a ready market available. Some urban farmers grow crops such as tomatoes, spinach and lettuce solely for commercial purposes but in general this is more common in western Africa than in eastern and southern Africa. Tree crops are not very commonly found due to the uncertainty of land tenure that many urban farmers experience.

Urban farmers face various constraints such as irregular rainfall, drought, flooding, water logging, poor soils, pests and disease, and the destruction of crops by animals, all of which are no different from the problems faced by rural farmers. Other problems, however, are more specifically related to the urban context and particularly confront the poor who practise off-plot farming. Examples include uncertainty regarding land tenure, theft of crops, lack of capital and inputs, the threat of eviction and the possible destruction of crops.

In many African countries, urban farming is illegal. By-laws frequently date from colonial times and forbid all agricultural activity within the boundaries of urban centres. However, as the practice has become increasingly widespread over the last two decades, a change in policy has occurred. During the 1960s and 1970s, policies were restrictive in the sense that harassment and destroying of crops were common measures taken by the local authorities. In the 1980s, however, a gradual shift in attitude took place and nowadays, urban farming is usually tolerated as long as it does not become a nuisance. As far as crop cultivation is concerned, the height of a crop, particularly maize, is important because it is said that criminals can hide in it and mosquitoes are assumed to breed in the axils. In some urban centres, for example Dar es Salaam, the local authorities are encouraging the practice of urban farming in order to raise food-supply levels.

Urban agriculture is considered by many as an environmental hazard. Livestock can cause noise, bad smells, traffic accidents (when roaming in the streets) and spread diseases. Crop cultivation can cause soil erosion, contaminated water can be used for irrigation purposes and crops cultivated along the road sides are prone to air pollution. Since urban farming tends to be more intensive than rural farming, the use of chemical fertilisers, pesticides and insecticides can have an impact on the urban environment, causing pollution in not only the plants but also the soil and groundwater. The recycling of sewage and urban solid waste and turning it into compost is often put forward as a kind of panacea for both urban crop production and the improvement of the urban environment. Although environmental awareness is growing in Africa, such measures have not (yet) been put into practice.

Urban agriculture in Kenya

The first and up to now most comprehensive study on urban agriculture in Kenya was the one carried out by the Mazingira Institute in 1985 (Lee Smith *et al.* 1987; Memon & Lee-Smith 1993; Lee-Smith & Memon 1994). The study was carried out in six towns of various sizes (including Nairobi) thought to be representative of 'urban Kenya' as a whole. The study population consisted of households from all income categories. It was found that farming is a very common activity among urban households: almost two-thirds grew part of their food; 29% of the urban households did so within the boundaries of the town in which they lived (i.e. urban agriculture per se). Almost half of the households kept animals; 17% did so within the town boundaries. It was estimated that about 25 million kg of crops were produced in Kenya's urban areas in one season and some 1.4 million animals were kept. Most of the agricultural produce – both crops and animals – was meant for subsistence purposes, which is related to the fact that most urban farmers appeared to be women and that most households carrying out urban farming belonged to the lower income categories.

All other studies regarding urban agriculture in Kenya focused on the country's capital, Nairobi. Freeman undertook a survey in nine randomly selected open spaces in Nairobi in 1987, using plots to locate respondents (Freeman 1991, 1993; Lado 1990). Mwangi as well as Dennery concentrated on Nairobi's poor, the former studying the importance of urban farming for the households' food security and nutritional condition, the latter focusing on decision-making among food producers (Mwangi 1995; Mwangi & Foeken 1996; Foeken & Mwangi 1998; Dennery 1995, 1996). What is common in all these studies is not only the widespread occurrence of food production within a city like Nairobi, but also the practice of 'traditional' farming systems and techniques, the importance of urban food production as a source of both food and income, and the constraints the producers face. As for the economic value of urban farming, it is especially important for the low-income groups, and female-headed households in particular. Regarding the constraints, pests and diseases, theft and lack of access to land are the prevailing problems.

Four different farming systems can be distinguished in Nairobi (and in most other Kenyan urban centres as well) (Foeken & Mwangi 2000). The first one, *small-scale subsistence crop production*, is by far the most common one, particularly among the low-income households. Farmers always plant a variety of crops on their *shambas* (plots), but basic staples like maize, beans and kale (*sukuma wiki*[5]) particularly stand out as the crops cultivated by the large majority of the farmers. Conspicuously absent are tree crops, for reasons of limited space (many plots are too small) and uncertainty regarding land tenure. The labour needed is mainly done by women. For instance, in 80-85% of the farming households in the low-income areas of Korogocho and Pumwani/Eastleigh, the women were responsible for the farming activities (Mwangi 1995). Cultivation practices are usually very simple: the *panga* (sturdy bush knife) and *jembe* (hoe) are about the only tools used. The use of 'modern inputs' is quite limited. Maintaining or improving soil fertility is mainly done by means of animal droppings or other organic material. Only a (small) minority of the farmers uses chemical inputs, because most farmers cannot afford it.

The second farming system concerns *small-scale market-oriented crop production.* Despite its potential in terms of food, employment and income, small-scale crop production entirely for commercial purposes is a rare phenomenon in Nairobi. In areas like Dagoretti and Kasarani very small farms (0.25-2 acres) exist where vegetables are produced for the urban market under intensive production systems (Mugambi 2002). Most of these farms are located at valley bottoms where irrigation is possible the whole year through. Besides this type of small-scale commercial farming, a few other types exist. The first one concerns ornamental crops, grown in plastic bags. It is commonly more well-to-do people who engage in this activity and who have employees to run the place. The plants are mainly seedlings sold to individuals and landscaping companies. The second type also concerns seedlings, notably of vegetables,

[5] *Sukuma wiki* is a typical ingredient in the diet of the poor households, favoured as the usual supplement with the basic *ugali* dish (stiff maize porridge). It grows fast, gives high yields, and has a high nutritional value.

grown on very small plots. An example is the Mathare Self-Help Group consisting of jobless slum dwellers. The group succeeded in obtaining permission from the City Council to till land next to the road in Kariokor. The seedlings are sold to farmers as far as the rural areas of Kiambu. Finally, Freeman (1991) mentions a very special crop: 'natural hay'. He noticed that Kikuyu women scythed the lush grass on roadside verges with their *pangas*, to be collected by dealers for selling on the market as animal fodder. Although not a cultivated crop in the strict sense, Freeman considers the crop as "a product of the city's open spaces with evident commercial value" (p. 92).

The third farming system, *small-scale livestock production*, is often combined with subsistence crop cultivation, but is usually for both subsistence and commercial purposes. Livestock is a quite common sight in Nairobi (and other Kenyan towns), especially in the open spaces in the outskirts of the city. Nevertheless, zero-grazing is the dominant production system (Mukui 2002). Particularly among the urban cultivators, livestock keeping is quite common: Freeman (1991) found that over half of 'his' cultivators kept some livestock. Poultry is by far the most common species. Recent figures from the Ministry of Agriculture[6] show that there were some 68,000 layers and 290,000 broilers in Nairobi in 2001 (Isika *et al.* 2002). Moreover, the population of pigs was estimated at about 41,000, sheep and goats at 40,000 (1999 figure), dairy cattle at 17,000, indigenous cattle (zebu) at 6,000, and rabbits at 9,000. About 28.6 million litres of milk were produced in Nairobi in the year 2000. If space were available, many more people would like to keep livestock. There is little knowledge on inputs for livestock rearing. Practices like dipping, spraying, vaccinating and using veterinary drugs are not very common. This partly explains the high mortality rate among the Nairobi livestock. Most farmers give additional feeding to their animals, such as crop residues and/or urban waste.

Finally, *large-scale commercial farming* can be found in the south-western part of the city (Mugambi 2002). These are mainly dairy farms – with 20 to 100 dairy cows – that sell their milk to processors or large hotels. Three flower farms are located in the Karen/Langata area. These flowers are exported by plane to the Aalsmeer flower auction in The Netherlands. Finally, a number of farms are specialized in raising pedigree cattle and racing horses.

---

[6] Because the City of Nairobi has provincial status as well, figures for Nairobi are available in the annual provincial reports of the ministry, unlike other cities and towns in Kenya.

Aspects of urban farming

In this section, five aspects of urban farming are briefly discussed: socio-economic aspects, farming techniques, environmental issues, the legal and institutional setting, and urban-rural linkages. The first three of these aspects form the core of this book. The fourth aspect (legal and institutional setting) is here and there dealt with in the context of the other aspects. Finally, although the fifth aspect (urban-rural linkages) is certainly also of importance in relation to the urban dwellers' food provision, this aspect is only touched upon, because it forms the topic of a separate study (Owuor 2003, 2006; Foeken & Owuor 2001).

Socio-economic aspects

In socio-economic terms, urban agriculture can be caught in three, interrelated aspects: gender, income and food security. As for gender, in general, urban agriculture is dominated by women, at least in eastern and southern Africa.[7] This applies in particular to the low-income urban households for whom urban farming is done either for raising the household's food security or for survival. In Kampala, the 'survival' group appeared to consist mainly of female-headed households (Maxwell 1994). Traditionally, providing the household with food is the women's responsibility. This may also explain the fact that it is mainly staple food crops that are cultivated.

Urban agriculture is attributed a potentially beneficial role in terms of the urban economy, urban food supply and urban development in general (Smit *et al.* 1996). Although largely an informal economic activity, urban farming provides employment as well as an income for those involved. This income can be directly realised through the sale of crops or indirectly as a result of the need to purchase less food. Studies from Tanzania revealed substantial direct incomes from selling crops (Stevenson *et al.* 1996; Kiango & Likoko 1996) and even more so from selling livestock products, milk in particular (Sawio 1993; Mlozi 1997; Foeken *et al.* 2004). However, the impact on income is usually the indirect one, which is indicated by the term 'fungible income' (Smit *et al.* 1996), i.e. money otherwise spent on food is saved so that other necessary expenditures can be done. Evidence of this indirect impact was found by Maxwell (1995) in Kampala and Mwangi (1995) in Nairobi. In Dar es Salaam, a group of home gardeners was reported to save a sizeable amount of money each month by producing part of their own food (Mlozi 1998).

---

7 See e.g. Rogerson 1992; Lee-Smith & Memon 1994; Sawio 1994; Streiffeler 1994; Tinker 1994; Matshalaga 1996.

At the town or city level, urban farming contributes positively to the provision of fresh food (horticulture, fruit, eggs, milk, etc.) for the urban dwellers. However, this contribution varies from city to city, ranging from 20% in Windhoek (Namibia) and Gaborone (Botswana), 30% in Lilongwe and Blantyre (Malawi), to 50% in Nampula (Mozambique) and 50-90% (depending on the type of vegetables) in Dar es Salaam (Egal *et al.* 2001). But because of its usually low productivity, the sector's potential in terms of food supply as well as employment is much higher than presently appreciated, as various studies have indicated (for an overview, see Nugent 2000).

Food producers in town, especially those in vulnerable groups, benefit directly in terms of increased food security (Armar-Klemesu 2000). However, in the few studies comparing the energy intake of producers and non-producers, the former's intake level was usually only slightly higher than that of the latter (Egal *et al.* 2001). In a slum area in Nairobi, Mwangi (1995) found that farming households were better off in terms of energy consumption when compared with non-farming households. Although the difference was not statistically significant, the farmers benefited also in another way, namely by being less dependent on food gifts and food transfers during periods of relative food scarcity. Moreover, growing food also helps improve the quality of people's diets by providing fresh fruit and vegetables. Defining food security as "secure access at all times to sufficient food" (Maxwell & Frankenberger 1992: 8) implies that in the longer run, access to urban land is certainly a factor when talking of household food security. As Maxwell (1995: 1677) points out, losing access to urban land is much more of a risk in this respect than the actual amount of land.

Equally little is known about the nutritional impact of urban farming activities for the producers. In the same study by Mwangi (1995), it was found that wasting (low weight-for-height, indicating acute malnutrition) was lower among under-fives in producing households, while for stunting (low height-for-age, indicating chronic malnutrition) no differences were found. In Kampala (Uganda), it was the other way around, stunting being significantly lower among children in producing households compared with non-producing households, while the wasting levels were about the same (Maxwell 1995). Also in Kampala, Semwanga (2002) reported a higher probability of having malnourished children in households not owning livestock than in households with livestock, be it that the difference was, again, rather small.

Farming techniques

Studies so far carried out in Eastern Africa revealed that farming systems among the urban farmers are not different from the systems practised in the rural areas where most of them came from.[8] Most households grow staples like maize, while vegetables are also very common. This is not surprising since many of the urban farmers are poor women whose prime responsibility is to feed the household.

Since most plots are very small, farming techniques are very simple. The only tools used are usually hoes and cutlasses (Rakodi 1995; Gbadegesin 1991). Moreover, inputs like chemical fertilizers, insecticides and pesticides for crop cultivation are generally used sparsely (e.g. Gbadegesin 1991; Lee-Smith & Memon 1994; Mwangi 1995), one exception being the crop cultivators in the Tanzanian town of Mbeya[9] (Foeken *et al.* 2004). The reason for this may be twofold. First, households are often too poor to be able to afford such inputs. Secondly, since the land that is cultivated by the urban poor often does not belong to the ones who cultivate it, any investment is too risky. However, it may be expected that those urbanites who farm on their own compound do use 'modern' inputs.

Except for those who use their backyard for farming purposes, irrigation is not common. In Mbeya and Morogoro (Tanzania), about one in ten crop cultivators irrigated their crops (Foeken *et al.* 2004). In Nairobi, Freeman (1991) came across one out of each eight cultivators practising some kind of irrigation. Cornish (2002a) estimated that an area of 2,220 hectares within a radius of 20 km from the city centre was under irrigation. For many of the poorer farmers, only those who have plots along a river can benefit from the yearly flooding of the river bringing water and nutrients into the soil. This water is highly polluted, however, hence the crops grown on these fields can be harmful for humans. Irrigation with sewage water is not uncommon. For instance, in Kibera, about 25% of the farmers used it (Dennery 1995). According to Cornish (2002b), the use of sewage water has increased during the past ten years. Although tapping from sewage pipelines is illegal and serious health risks are involved (faecal contamination), the water and sewage authority is aware of the practice but seems to advocate a policy of what Cornish calls "grudging acceptance".

Urban farming and the urban environment

Urban agriculture is considered by many as an environmental hazard because of the danger of soil erosion and the use of contaminated water for irrigation purposes, while crops cultivated along road sides are prone to air pollution

---

[8] See e.g. Lee-Smith *et al.* 1987; Freeman 1991; Egziabher *et al.* 1994; Maxwell 1995; Mwangi 1995.

[9] Where one-third of the cultivators used chemical pesticides, 40% insecticides and three-quarters chemical fertilizers. Interestingly, the percentages of farmers in a Tanzanian town of comparable size (and also comparable with Nakuru), Morogoro, using chemical inputs were much lower (Foeken *et al.* 2004: 58).

(Mosha 1991). Since urban farming tends to be more intensive than rural farming, the use of chemical fertilizers, pesticides and insecticides can have a great impact on the urban environment. Animals can cause overgrazing and traffic accidents (Smit *et al.* 1996: 205-206). At the same time, it is the urban poor who are most at risk in relation to the environmental problems of the cities, which were in the mid-1990s described as being 'very significant' (Burgess *et al.* 1996: 19) and as having reached 'crisis proportions' (Nuwagaba 1996: 23). It is the poor who suffer most from lack of access to land for purposes of housing and economic activities, from lack of access to water for both drinking and farming, from the effects of air pollution for both people and crops, and from the dumping of the city's waste often in the vicinity of the slum areas. In this context, if well-managed and legalized, urban agriculture is seen as a solution for both lessening the food insecurity of the urban poor and improving the urban environment in general (UNCHS 1996: 410-411; Brock & Foeken 2005). Or, to put it differently, "urban farming can help to create an improved micro-climate and to conserve soils, to minimise waste in cities and to improve nutrient recycling, and to improve water management, biodiversity, the O_2-CO_2 balance, and the environmental awareness of city inhabitants" (Deelstra & Girardet 2000: 47). Land use, use of (partly treated) sewage water, use of composted solid waste as well as other types of organic waste (plants, animals) are all components of the so-called 'closed-loop system' that existed in the pre-industrial city: "liquid and solid city wastes are returned to the land and serve as the prime source of soil building and enrichment for the production of perishable food for the city" (Smit *et al.* 1996: 12; Nelson 1996).

An attempt to change the attitudes of local governments was a Dutch initiative in Kenya called the Green Towns Project (Duchhart & Grootenhuis 1993). In the context of this programme, local authorities receive training in urban planning, with special emphasis on the integration of environmental issues in the Local Authorities Development Programmes. In this approach, proper urban agriculture is implicitly part of sustainable urban development. Three Kenyan towns had been chosen as pilot towns, namely Eldoret, Nanyuki and Migori.

Legal and institutional setting

Crop cultivation and livestock keeping in town is practised under different conditions than farming in the rural areas. It takes place "within the confines of residential areas and in a legal regime (...) that restrict activities associated with both crop and livestock production" (Mukui 2002: 9). In Kenya, the Local Government Act, Water Act, Public Health Act, Animal Diseases Act, as well as several other acts, provide each local government with the power to restrict farming activities in town. In practice, it means that although officially forbidden, city authorities tolerate urban farming (Gathuru 1993). Still, in the mid-

1980s different local authorities in Kenya pursued different policies regarding urban farming. For instance, in Nairobi, Mombasa and Kakamega, a laissez-faire attitude (or neglect) prevailed, while the local governments in Isiolo and Kitale implemented favourable measures for urban farmers (Memon & Lee-Smith 1993).

Open space planning in Nairobi is administered by zoning regulations based on the 'city beautiful' idea dating from the colonial period. Hence, urban agriculture was long considered a misfit in urban development (Mwangi 2002). For instance, it was seen as a traditional activity carried out by recent immigrants from the rural areas. Studies have shown that this is a myth: urban farmers are those who have been in town for quite a number of years; are settled so to say (see e.g. Mougeot 1993; for Kenya: Mwangi 1995; Mwangi & Foeken 1996). Urban farming was also seen as a subsistence kind of economic activity, incompatible with the 'modern' character of the city. Through the years, zoning regulations have changed somewhat, particularly regarding informal sector activities. With written permission – a so-called temporary occupation licence or TOL – livestock may be grazed on the outskirts of the city (Munari 1994; Karanja 1994). The regulations regarding crop cultivation have not changed and this activity is still officially forbidden. However, there do exist zones where agriculture is permitted (Quon 1999), notably on the outskirts of the city, i.e. the areas that came to be located within the city boundaries after the expansion of 1964. Within the 'old' boundaries, the present policy is largely one of ignoring the activity, certainly as far as the urban poor in the informal settlements are concerned. In general, however, urban farming is not being incorporated in urban planning. In the most recent government policy initiative on urban planning (the Physical Planning Act of 1996), agriculture is still systematically excluded from the urban land use system, which, according to Mireri (2002) negatively affects the development of the sector.

The majority of the Nairobi farmers are completely left on their own, getting no assistance or advice of any sort. However, the Ministry of Agriculture does provide extension services, in principle to everyone who asks for it. Yet, roadside, riverside and sewage line farming are not recognized by the officers as it has been prohibited in the 1961 Nairobi City Council by-laws and never been reviewed since then (Ateka 1999). This implies that many of the poor urban cultivators do not qualify for extension. Moreover, those who do qualify are often either not reached or do not get the advice they need. According to Mugambi (2002), some of the reasons are that (1) the majority of the city farmers are part-time farmers, usually not at home during the extension officers' visiting hours, and (2) many farmers carry out specialized forms of agriculture, to which the advisory service is not adapted.

Very little research has been done on this aspect of urban agriculture. This is the more surprising as, according to the existing literature, some of the major problems faced by the urban farmers are related to it: security of access to land, too small plots, lack of infrastructure, harassment, lack of extension services, etc. Meanwhile, some NGOs have been trying to fill this 'gap' by actively assisting urban farmers. The Undugu Society's work in one of the Nairobi slums is a case in point (Gathuru 1993).

Urban-rural links

As far as rural-urban linkages in Sub-Sahara Africa are concerned, the focus has so far predominantly been on the urban dwellers contributing to the livelihood of the rural dwellers, usually by means of remittances of family members living in town. Hardly anything is known about the reverse flow, i.e. in how far urban households realise part of their livelihood from rural sources. In a survey in Harare in three residential areas of different socio-economic statuses, it was found that just over one-third of the respondents claimed to hold land outside the city from which they could receive food crops (Drakakis-Smith 1992). Two other surveys held in Harare in 1985 and 1988 revealed that respectively 40% and 53% of the households claimed to have access to rural land (Potts & Mutambirwa 1990). However, only about half of the 1985 population said they had used the land productively the previous year, despite the fact that the rains had been good. For those who actively farmed the rural land, the produce – either self-consumed or sold – represented a fairly significant addition to the households' income.

In a general survey of 1985 it was found that 55% of the Kenyan low-income urban population stated to have access to rural land (Lee-Smith *et al.* 1987), while at least one-third of them stated to have livestock back in the rural areas (Lee-Smith & Memon 1994). As for access to rural land, the same figure was found in 1994 among households in a Nairobi slum area (Mwangi 1995). Of the latter, 44% said to be the actual owners of the plot(s), while in all other cases parents or relatives appeared to be the owners. However, ownership by the urban households did not automatically mean that they also used the plot themselves: half of the rural plots owned by the urban households were either let to be used freely by others (mostly relatives) or were left idle. Further analysis of the 1994 data indicated that those of the urban poor who did have access to rural land were better off in terms of food security than those who did not (Foeken & Mwangi 1998).

Theoretical considerations

From a theoretical point of view, the study fits in the *livelihood* approach. A livelihood is defined as comprising "(...) the capabilities, assets (including both material and social resources) and activities required for a means of living" (Carney 1998: 4, derived from Chambers & Conway 1992). Central to the livelihood approach is that people should not be seen as passive victims of adverse circumstances,[10] but instead develop all kinds of actions and strategies aimed at preserving a certain livelihood level.[11] The keyword is 'access', the crucial question being in how far people have access to all kinds of resources (or 'assets' or 'capitals').[12] So important is 'access' that, according to Bebbington (1999: 2022), "[it is] perhaps the most critical resource of all", since, as Rakodi (2002b: 293) observes, "proximity and availability [of resources] mean little if access is denied".

The livelihood approach distinguishes five "vital" resources,[13] although their boundaries are not always that clear nor is the categorization exhaustive (Rakodi 2002a):

1) *Natural resources*: land, water, pastures, etc. Natural assets may be less significant in an urban setting (Meikle 2002), but with increasing reliance on agriculture (both urban and rural), access to land and security of tenure have become important resources for urban dwellers (see Payne 2002).

2) *Physical resources*: basic infrastructure and services (shelter, transport, water, energy, communications, hospitals), equipment, tools, inputs, food stocks, household assets, livestock, etc.

3) *Financial resources*: savings, loans, credit, wages/salaries, pensions and remittances. Urban households are highly monetized and so access to a monetary income is essential.

4) *Human resources*: capabilities, skills, experience, labour, knowledge, creativity, health, etc. These are important to the fulfilment of productive and reproductive tasks. Capacity to work is the main asset of the urban poor. Lack of skills and education affects the ability to secure a livelihood in towns more directly than it does in the rural areas.

---

10 I.e., the adverse circumstances as briefly described in the first section of this chapter.

11 See e.g. Chambers 1983; Jones 1999; Rakodi 2002a; De Haan & Zoomers 2003; Kaag *et al.* 2004.

12 For more details on these types of resources in an (African) urban setting, see for instance Rakodi 2002a and Brown & Lloyd-Jones 2002.

13 See e.g. Mitlin 2003; De Haan 2000; Rakodi 2002a; Carney 1999; Chambers 1995; Blaikie *et al.* 1994; and Chambers & Conway 1992. Some scholars use the term 'assets' or 'capitals' instead of 'resources'; here, the terms 'assets' and 'resources' are used interchangeably.

5) *Social resources*: formal and informal networks from which various opportunities and benefits can be drawn by people in their pursuit of livelihoods. These are mainly reciprocity and trust embedded in social relations, social structures and societal institutional arrangements. Closely linked to social resources are *political resources* based on access to the political process and decision-making (see Devas 2002). Meikle (2002: 42) elaborates that the urban poor are linked into structures of governance through their dependence on or exclusion from the delivery of infrastructure and services by municipal authorities.

Although the livelihood approach distinguishes these five types of resources, the importance of "cultural resources" in livelihood studies should also be recognized. Such cultural aspects as language, taboos, cultural institutions, religion, etc, may have an important influence on an individual's or a household's livelihood strategy.

Returning to the topic of this book, urban agriculture, many of the above resources form a necessary condition for a household or individual to be able to undertake urban farming.[14] *Natural resources* refer to land (both quantity and quality), water and energy, etc. Access to land is an important prerequisite without which it is impossible to practise urban farming.[15] Particularly for the urban poor who often lack a compound, *public* space is a crucial resource for farming in town (Brown & Lloyd-Jones 2002). *Physical* or *productive resources* relate to farming tools, inputs for both crop cultivation and livestock keeping, but also to, for example, food stocks (for animals). *Financial resources* include money (either at home or in the bank), a loan or credit but also income-generating activities, in order to be able to pay for inputs (including hired labour) or to invest in e.g. livestock. *Human resources* concern not only the household's own labour as such, but also the quality of it in relation to farming. Finally, *social resources* include, for instance, local networks between urban farmers (e.g. cooperatives, women's groups) and relations between sellers and buyers.

Access to these resources is included in the various aspects of urban farming dealt with in this book and briefly discussed in the previous section. Especially the socio-economic situation of the household can be regarded as a major condition for getting access to all kinds of resources, household income and gender being particularly relevant (Kaag *et al.* 2004; Beall 2002). For instance,

---

14 It should be stressed that the book does not deal with the overall livelihood of households, of which urban agriculture is one element (as done, for instance, by Martin *et al.* 2000 and, for some specific households in Nakuru, by Owuor & Foeken 2006). Instead, the focus is on urban farming itself (being part of a household's livelihood) and the resources needed to engage in this activity.

15 Although from the literature there are examples of people growing crops or keeping animals on the balcony, the roof or in a room in the house (Smit *et al.* 1996).

poor households and women have less easy access to land and can less easily obtain title deeds ('legal and institutional setting'), while they are less inclined to use modern farming techniques and/or inputs ('farming systems and techniques'). Moreover, they are less likely to have access to rural land, and if they do, have greater difficulties in using it productively because of the higher costs involved to get there.

In theory, the legal and institutional setting, including national policies, local by-laws and local policies (see e.g. Foeken *et al.* 2004; Foeken 2005), could be a major conditional factor for getting access to various resources as well. For instance, local by-laws contain such regulations as to where urban farming is allowed, the types of crops one may cultivate, as well as the number of certain types of animals (if at all) allowed to keep and how to keep them. Moreover, local authorities can at will decide to follow a policy line of repression, tolerance or promotion. However, in practice, a laissez-faire policy ('toleration') has prevailed in sub-Saharan Africa during the last decades, even though there are numerous examples of harassment and destruction of crops. In other words, the relevance of the legal and institutional setting as a regulating force in relation to getting access to resources for urban farming has mostly been more of a paper than a practical matter.

The previous discussion implies that some people do *not* have access to certain types of resources necessary to engage in urban agriculture (besides those who simply do not want to farm in town). This brings us to the concept of exclusion. As Kaag *et al.* (2004: 61) rightly observe, "not all members of a given society have equal access to social security arrangements: particularly the poorest tend to be excluded". For households facing exclusion – or 'denied access' – to such resources, farming in town as a coping strategy in order to sustain one's livelihood is no option. For those, access to rural land and/or social resources might be another possibility to fulfil one's needs.

The term exclusion comes close to the concept of 'entitlement' introduced by Sen (see for instance Sen 1981). The main difference with exclusion is that entitlement refers to the *right* of access, while exclusion and by definition also 'inclusion' concerns access itself, i.e. the question of whether the entitlement can be effected into access to specific resources. The importance of Sen's concept lies particularly in the notion that the right of access is largely determined by factors belonging to broader social, cultural, economic, political and natural spheres. For instance, occupying land belonging to someone else is forbidden, as is usually growing maize in town ('legal setting'). Hence, one has no right to grow maize on a plot to which the cultivator has no right; in other words, the cultivator violates the law in two ways. The fact that these practices do occur, however, shows that people can have access to resources even without the right to do so. This says something about the difficult situation many people find themselves and about the way the legal systems in many African countries work. It also shows that if local authorities would see to compliance with the law, people would be forced to look for alternative livelihood strategies.

Two more notions are important here. First, livelihood strategies are not static but change (Hoon *et al.* 1997). This can be due to changes in the (access to) resources at the disposal of the person(s) involved. It can also be because of changes in the context in which these persons are living. Second, resources are limited, so access to resources can be very competitive. This implies that access to resources (such as urban land for farming) for the one ('inclusion') may mean denied access for the other ('exclusion'). Or in the words of De Haan (2000: 26), "the sustainable livelihood of one actor may result in the social exclusion of another".

At another level, it is increasingly being recognized that urban farming is an element of the wider urban environment (and, hence, an aspect of urban management and urban development). Households engaged in farming activities within the town or city boundaries make use of urban resources such as land and water, but sometimes in a detrimental way. In order to make urban development 'sustainable', local authorities see it necessary to regulate and guide farming practices in town. In practice, this can easily lead to curtailing the possibilities for farming for the one, but also to more favourable conditions for another, for instance increased access to extension services or credit for the development of urban farming in designated zones. In other words, sustainable development at town or city level may lead to exclusion of certain resources (land, water) at household level. Not surprisingly, those with the least assets (the poor) are most likely the ones to suffer most from such developments.

The present study

The uniqueness of the study

Although there exists a certain body of knowledge regarding urban farming in sub-Saharan Africa (see Obudho & Foeken 1999), this knowledge is still very fragmentary and incomplete, because most studies focus on one or two aspects of urban farming only and mostly in one specific urban centre (usually the national capital) or even a specific part or project within that centre. In the mid-1990s, Mougeot (1994b) pointed out that particularly lacking are for instance studies in which urban farmers and non-farmers are compared,[16] studies in

[16] The Nairobi study by Mwangi (1995) is one of the few exceptions.

which different towns or cities are compared,[17] and studies in which the various aspects and effects of urban agriculture are analysed. Since then, very few studies have been done along one or more of these lines. The present study tries to some extent to fill this gap.

The study distinguishes itself from studies done so far because of the combination of three points of entry. First, different aspects of urban farming are integrated. Up to now, studies on urban agriculture focused on one or perhaps two aspects of the phenomenon. The present study covers to some extent the various aspects discussed above, with particular emphasis to the socio-economic and environmental aspects.[18] Moreover, an institutional form of urban agriculture, namely school farming in Nakuru town, is discussed in a separate chapter. The advantages of this integrative approach are twofold: (a) a comprehensive picture of urban farming and the urban farmers emerges, so that (b) the results of this study can be used for planning purposes.

Second, the study focuses on a medium-sized town: Nakuru. Studies so far have almost without exception been done in the national capitals. However, a national capital is in many respects not representative for the whole urban population in a specific country. For instance Nairobi, being the national capital and being so much larger than any other urban centre in the country, is dominant in terms of economic, political and cultural aspects. As a result, the city attracts a continuously large flow of migrants from all parts of the country. A medium-sized town like Nakuru is much more common in sub-Saharan Africa than the large national capitals.

Third, there is special attention for the urban poor. Although the study does not focus solely on the poor, this category receives special attention. It is widely recognised that it is the urban poor who suffer most from the economic recession that has been going on since the 1980s. Due to the increasing rates of unemployment, they have to resort to the informal sector and/or fall back on farming in the city. For this group, all aspects of urban farming mentioned above come together in a negative way. They need it most, but at the same time are most prone to all kinds of constraints. For them, urban cultivation can at best be a temporary business in which it is too risky to invest. Hence, the benefits in terms of raising income and more food security are uncertain and limited. Yet, for many of them it is at the same time a way of survival.

---

17 A recent example is the study in Tanzania by Foeken *et al.* (2004) who compared two towns, Morogoro and Mbeya, with different physical circumstances.

18 The legal aspects are more extensively described elsewhere (Foeken 2005), while the urban-rural links formed a separate (PhD) study (see Owuor 2006).

Set-up of the book

The book is the reflection of a whole range of studies carried out in the context of the overall Nakuru Urban Agriculture Research Project (NUAP). The 'foundation' of the project was a general survey in 1999 meant to get a good general impression of the importance of urban farming in Nakuru town. The results of this survey are presented in Chapters 3 ('The farmers'), 4 ('The crops'), 5 ('The animals') and parts of Chapters 6 ('The benefits') and 7 ('The support'). Most of the poverty analysis in Chapter 10 ('The poor') is also based on this general survey.

A second survey was done one year after the first one, primarily aiming at the effect of urban farming for the food security situation of the households involved and for the nutritional condition of the household members. The results of this survey are mainly included in Chapters 6 ('The benefits') and 10 ('The poor'). The two surveys also offered good comparison possibilities between a 'normal' year in terms of rainfall and a dry year, which is used in various chapters.

Four sub-studies at Masters level were also part of NUAP. Two of these – one on the environmental impact of urban farming and the other on school farming – are laid down in separate chapters in this book, namely Chapters 8 ('The environment') and 9 ('The schools'). The results of a third study – on the impact of NGO support for urban farmers – are included in Chapter 7 ('The support'). The fourth sub-study was of an anthropological nature focusing on decision-making related to farming in town. The detailed findings of this study have been used throughout the book, mostly in the form of illustrations (citations).

The final Chapter 11 ('Summary and conclusions') tries to bring together the findings of the various studies, relating it to the theoretical considerations as outlined in Chapter 1 and translating it into policy recommendations.

Photo 1 Nakuru town seen from the lower slopes of the Menengai Crater. Lake Nakuru is in the left background. Maize cultivation in the foreground (peri-urban farming). (Dick Foeken, 1999)

Photo 2 Crop cultivation in the medium-density area of Bangla Desh. (Dick Foeken, 1999)

2

The setting

Nakuru town

Natural structure
Nakuru is located in the heart of the Great Rift Valley 160 km north-west of Nairobi between latitudes $0^{\circ}10'$ and $0^{\circ}20'$ South and longitudes 36° and $36^{\circ}10'$ East (see Map 2.1). The largest part of the town lies at an altitude of about 1700 metres above sea level. In the northern part, on the slopes of the Menengai Crater, the altitude rises to about 1850 metres. Nakuru is located in the midst of a concentration of geographical features that together make up the Lake Nakuru catchment basin. These include the Menengai Crater to the north, the Bahati Highlands to the northeast, the Eburu Hills and Lake Nakuru to the south and the Mau Escarpment to the southwest. The lake water catchment is served by several small, seasonal rivers, including the Enjoro and Ng'ossor which flow through the town. Due to its location on the floor of the Rift Valley with its volcanic soils, Nakuru is engulfed with whirlwinds of dust during the dry season, giving the town its name (*nakuru* means 'a place of winds' in the Maasai language).

With a 1949-2002 average annual rainfall of about 940 mm, Nakuru has a dry sub-humid equatorial climate. The 'normal' rainfall pattern throughout the year is indicated by the line in Figure 2.1.[1] The most important period of rain

---

1 The provision of the rainfall figures by Francis Mwaura and John Githaiga (University of Nairobi, Department of Geography) is gratefully acknowledged.

Map 2.1 Kenya and location of Nakuru town

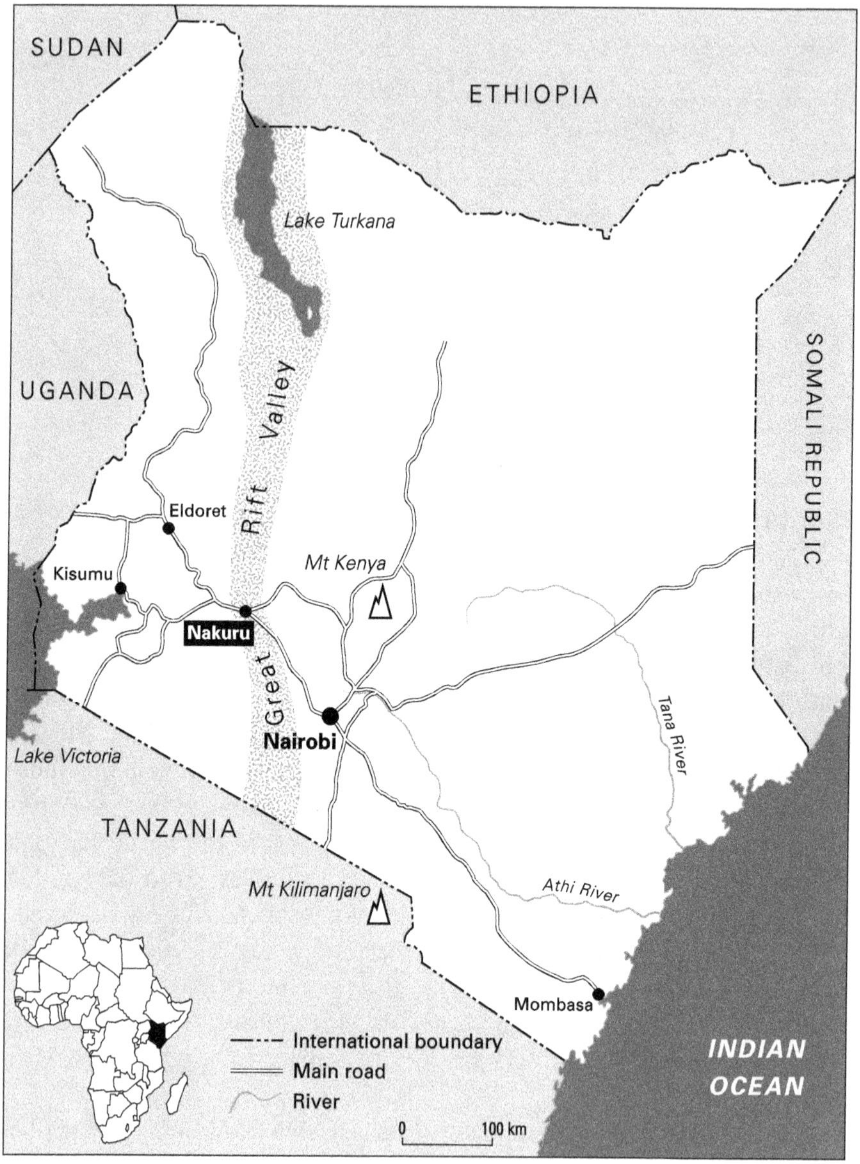

is the so-called 'long rains' during March-May, a period that coincides with the start of the growing season. Two minor rainfall peaks occur in July-August and around November. The latter period is called the 'short rains' during which some people try to plant a second crop if the long rains have failed. Figure 2.1 also shows the rainfall distributions of 1998 and 1999, i.e. the two years that were covered by the two surveys in this study (in 1999 and 2000, respectively; see below). The year 1998 was quite good in terms of rainfall. The distinct peaks in January and May account for the relatively high total rainfall in this year. More importantly, the distribution was more or less 'normal', resulting in 'normal' harvests as well. The year 1999, however, was quite different: after a promising start in March, the rest of the long rains largely failed, causing harvests to be relatively bad.

Figure 2.1 Monthly rainfall, selected years

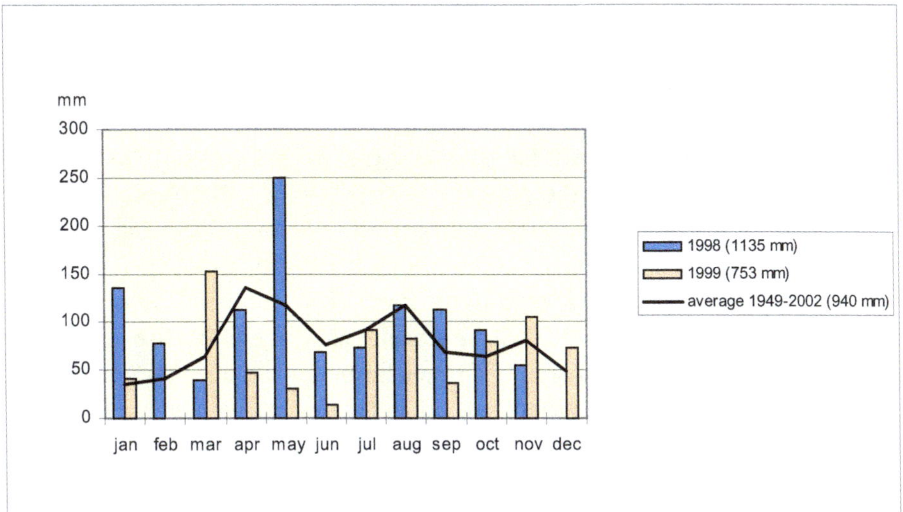

Historical development

Until the arrival of the railway at the beginning of the twentieth century, the present-day Nakuru area was used as grazing land by pastoral communities, mainly the Maasai (MCN 1999). Like Nairobi and Kisumu, Nakuru began (in 1904) as a railway station on the great East African Railway (or Uganda Railway) between the city-port of Mombasa on the Indian Ocean coast and Port Florence (today's Kisumu) on Lake Victoria. Being located in the so-called 'White Highlands' (the area of large farms owned by European settlers), Nakuru soon developed into an important regional trading centre and became the capital of the district with the same name and of Kenya's largest province, Rift Valley Province.

During the colonial period, Nakuru was a highly planned settlement, i.e. a square grid cut in two by the railway (De Meulder 1998; MCN 1999). The railway depots were north of the railway while the section south of the railway was the actual settlement, with its administrative, commercial and residential zones. The street pattern was as simple as it was efficient: streets with an east-west direction were called 'avenues' (numbered from 1 to 6) and streets with a north-south orientation were called 'roads'. Already during the 1920s, the town began to grow outside the original grid (Map 2.2). In the zoning plan of 1929 (the so-called Ballenden plan), Nakuru's further expansion was laid down, in accordance with the then generally accepted principles of functional zoning, i.e. with an industrial quarter, residential districts for the various social classes, a suitable location for a hospital and cemetery, recreational facilities, a site for the airfield, etc. One of the special residential quarters, to the southeast of the original grid, was Bondeni, meant for the Asian community. After the Second World War, and in particular after independence in 1963, public housing complexes were built for the African population. Thus, Nakuru was transformed from a colonial European town into an African town.

Since independence, Nakuru has known three major extensions of its boundaries, in 1963, in 1972 and the latest in 1992 (Map 2.2). The present built-up area coincides largely with the 1972 boundary (see below). With the extension of 1992, Lake Nakuru National Park fell within the municipal boundaries, as well as a stretch of agricultural land to the northwest of the park (and a narrow strip bordering the northeastern boundary of the park, which is also an agricultural area). Due to the subdivision of former farms into small plots for residential use, this stretch is now a largely sub-urban area, albeit with a strong agricultural character. Another sub-urban area extends to the north of the town, mainly on the slopes of the crater. Although this area falls outside the municipal boundaries, it is part of the Nakuru planning area (or the Nakuru Metropolitan Area as it is called). The total area of the municipality is about 300 km^2, of which the lake takes up 40 km^2. Due to the fact that the town is squeezed between the Menengai Crater to the north and Lake Nakuru National Park to the south, the present expansion is mainly to the east and the west, giving the town its elongated, east-westerly shape.

Population growth

Over the past 30 years, the population of Nakuru town has increased by a factor five. In 1969, the population was 47,151 (Kenya 1970), increasing to 92,851 in 1979 (Kenya 1981) and 163,982 in 1989 (Kenya 1994). Today, Nakuru is the

Map 2.2 Evolution of the boundaries of Nakuru town
(From MCN 1999, Fig. 3.4, p. 31)

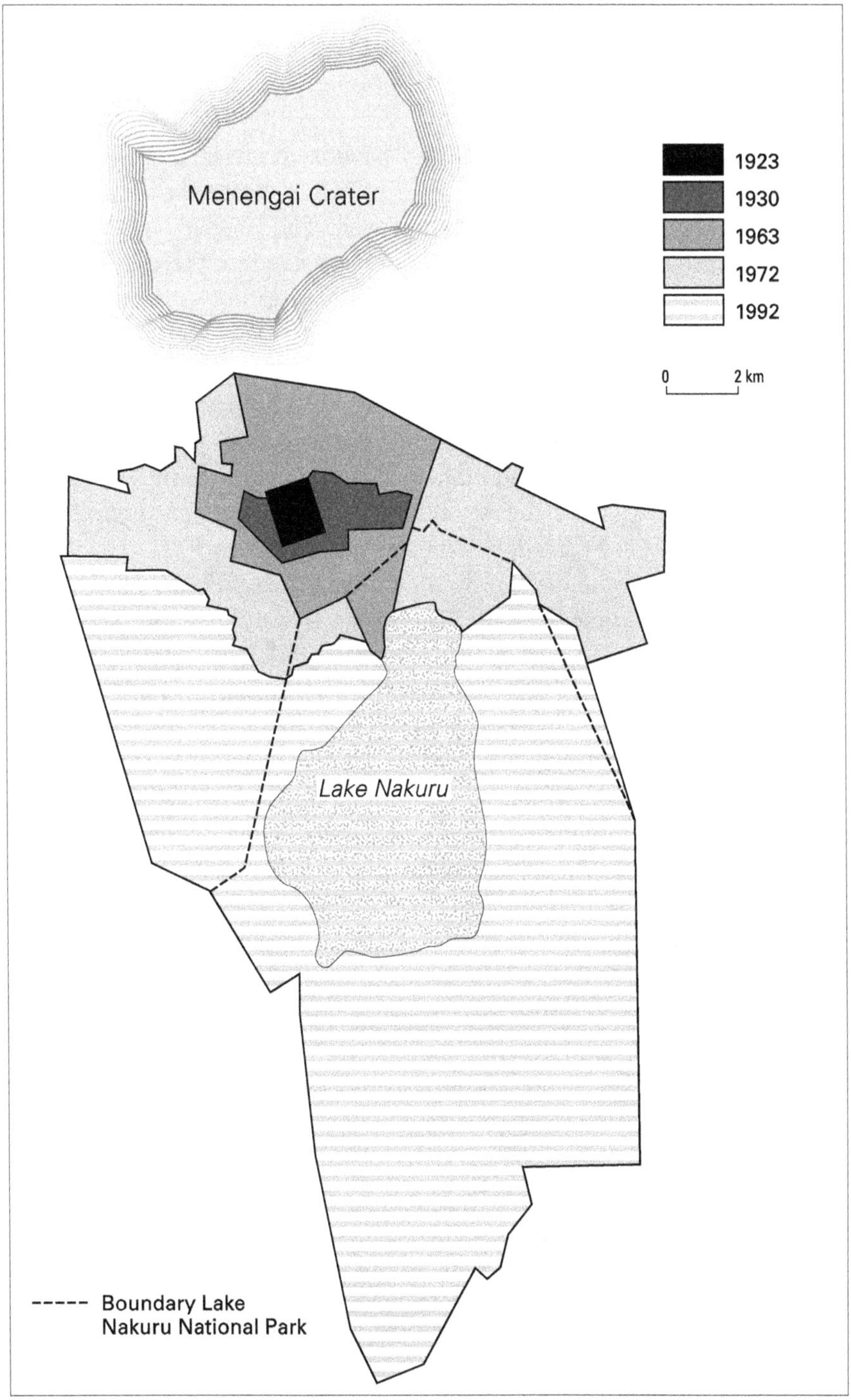

fourth largest town in Kenya (after Nairobi, Mombasa and Kisumu), with a population of 239,000 in 1999 (Kenya 2000). Intercensal annual growth seems to have continuously declined since 1969, namely from 7.8% between 1969 and 1979 to 6.5% between 1979 and 1989 and to 4.3% between 1989 and 1999. The annual growth figure of 7.8% for the 1969-79 period may be misleading, however, due to the boundary extension of 1972 (see Map 2.2), so that the real growth rate during this period may actually have been lower. In 1992, there was another boundary extension, so the 4.3% growth rate during the 1989-99 period may to some extent be an overestimation as well. The conclusion is that despite the substantial growth of the absolute Nakuru population, the growth rate has decreased quite rapidly during the last two or three decades.

Urban poverty

In 2001, the Kenyan government defined 'poverty' as inadequacy of income, deprivation of basic needs and rights, and lack of access to productive assets as well as to social infrastructure and markets. In quantitative terms, the 'absolute poverty line' is the income needed to obtain basic food and non-food items. For urban areas, this was Ksh. 2,648 per person per month in 1997 (Kenya 2001: 11), which was equal to the official minimum wage as set by the government at that time.[2] According to the Welfare Monitoring Survey that was held in 1997 (WMS III), 26,378 (or 41%) of the households in Nakuru Municipality were living below the absolute poverty line (Kenya 2001). Since poorer households tend to be larger than better-off households, the percentage of the Nakuru population affected was even higher, almost 50%.[3] Importantly, the 1997 figure implied a substantial increase in the prevalence of urban poverty compared with three years before when the number of households below the absolute poverty line stood at 30% (Ibid). This is related to the fact that "only a fraction of the [Nakuru] labour force is actually employed" (MCN 1999: 62). As a result, "there is a high dependency ratio, increasing unemployment and increasing urban poverty" (Ibid).

Economic structure

Nakuru's economy is based on commerce, industry, tourism, agriculture and tertiary services. Commercial activities are concentrated in the original Central

---

2 On 1 May 2002, the minimum wage was raised to Ksh. 3,500 (about US$ 50), so the number of people below the poverty line increased automatically as these official measures usually have little impact on the wages paid by employers to their employees.

3 Calculated as follows: according to WMS III, 113,674 individuals were affected, which was 49.4% of an estimated total 1997 population of 230,000 (based on the population figures of 1989 and 1999).

Business District (CBD), along various strips and in several smaller nodes (Map 2.3). Informal commercial activities have become an increasingly common feature of the town. Small-scale business and hawking activities are mainly to be found at major transport termini and on the reserves of busy internal roads (MCN 1999).

Because of the rich agricultural hinterland, Nakuru is called the "farmers' capital" of Kenya and is famous for its agro-based industries. There are over 100 agro-industrial establishments ranging from food processing to farm machinery assembly (MCN 1999). The main industrial zone is located west of the CBD (Map 2.3). More recently, a second industrial zone has developed to the east, with all kinds of related urban development activities.

There are several tourist attractions in and around the town. Of these, Lake Nakuru National Park is by far the most important, attracting visitors from all over the world. Minor attractions include the Menengai Crater and two archaeological sites. Furthermore, Nakuru offers a central point of departure to other attractions in the Rift Valley region, such as Lake Bogoria and Lake Baringo.

Besides being the "farmers' capital", there is large- and small-scale farming within the boundaries of the municipality. Large farms can be found in the west (Map 2.3), including the huge farm belonging to the Rift Valley Institute of Science and Technology (RVIST). Small-scale farming activities are developing within the municipality (MCN 1999) and are mostly located in the peri-urban areas. The former rural area south of the Enjoro River in the southwestern part of the town, which became part of the municipality after the boundary extension of 1992 (see Map 2.2), is one such area. Many farms have been subdivided into small-holder parcels and urban residential plots. Nevertheless, farming is still the main activity there.[4]

In addition to these economic activities, Nakuru town is an important transport and administrative centre. The 'rail-road ribbon' of both the Mombasa-Nairobi-Kisumu/Uganda railway and the Mombasa-Nairobi-Eldoret/Kisumu/-Uganda road runs through the centre of the town. This has attracted all kinds of support facilities, such as petrol stations. The town is also an important administrative centre. Being the capital of Kenya's most populous district – Nakuru District with a population of 1.2 million in 1999 – and the country's largest province – Rift Valley Province with a 1999 population of 7 million (Kenya 2000) – the town has a wide range of offices offering employment in the administrative sector to many people.

---

4 More detailed information on farming in Nakuru town is presented below in this chapter.

Map 2.3 Economic structure (from MCN 1999, Fig. 3.9, p. 45)

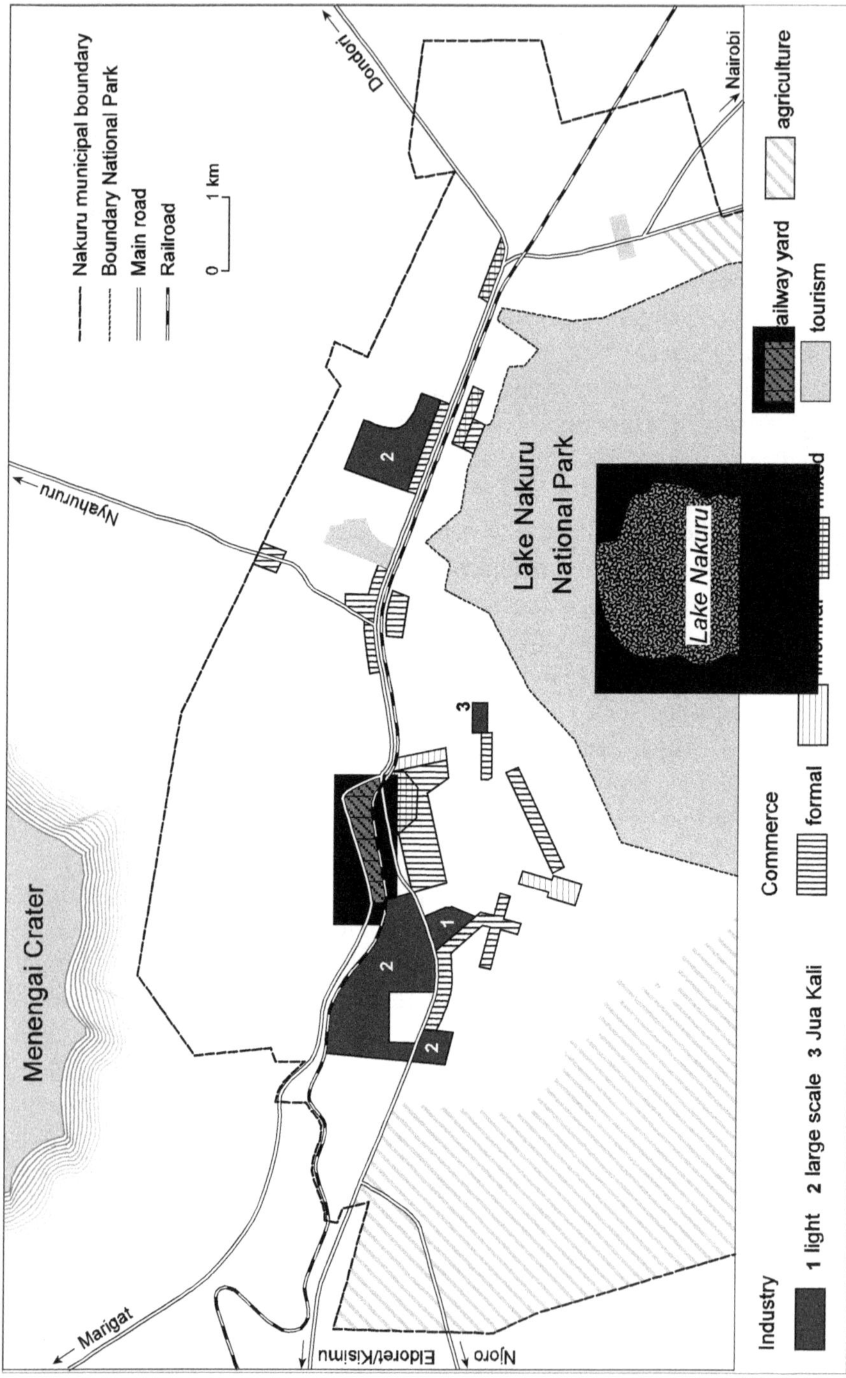

Land tenure and settlement structure
During pre-colonial times, land was communally owned. Nowadays, it is all in public or private ownership. Public land is owned by either the municipal council or the central government and is used for municipal and government purposes or leased out for a specified period to individuals for pre-determined urban land-use activities (MCN 1999). Public land constitutes the bulk of the municipal area. Except for the above-mentioned area south of the Enjoro River, which is private land, nearly all residential estates are leased public land.

Urban farming is highly dependent on the availability of space. In other words, housing density more than population density determines whether farming in a certain residential estate is possible and to what extent. Most of the low-density housing areas are located north of the rail-road ribbon, while south of that line medium and high-density housing areas prevail (Map 2.4). To some extent, housing densities coincide with income levels, in the sense that high-income areas generally have low densities (such as Milimani in the north) while low-income areas have high densities (for example Kwa Rhonda in the southwest). However, low-income areas with quite low housing densities also exist, examples being Ziwani, Flamingo and Kivumbini estates.

Environmental infrastructure
Sewage disposal in Nakuru town is by sewer reticulation, septic tanks and cesspools and pit latrines. There are two sewage treatment plants (indicated with an 'S' on Map A2.1, p. 176), the Old Town treatment plant within the boundaries of Lake Nakuru National Park and the newer and bigger Njoro treatment plant in the southwest, both using stabilization ponds as treatment mechanisms (MCN 1999). The capacity of the sewage system is under-utilized, mainly because of the inadequate sewage network (less than 20% of the built-up area is served by it). According to an informant from the Municipal Council, the under-utilization of the Old Town treatment plant is partly caused by the fact that people living just north of the park boundary use sewage water for irrigation purposes. This is an area (Block 14) with a lot of open space that is intensively used for crop cultivation.

In Nakuru, a lot of solid waste is generated from household, commercial and industrial activities. However, waste collection and disposal services are totally inadequate and are limited to the old town (MCN 1999). In many residential areas, waste collection relies on private initiatives, including some non-governmental and community-based organizations. There is one designated dumping site, a natural ravine, in the northwest of the town where waste is dumped without any form of separation. Where the ravine has been filled, the garbage has been covered with a thin layer of soil in which food crops are now being cultivated.

Map 2.4 Structure of housing (From MCN 1999, Fig. 3.6, p. 37)

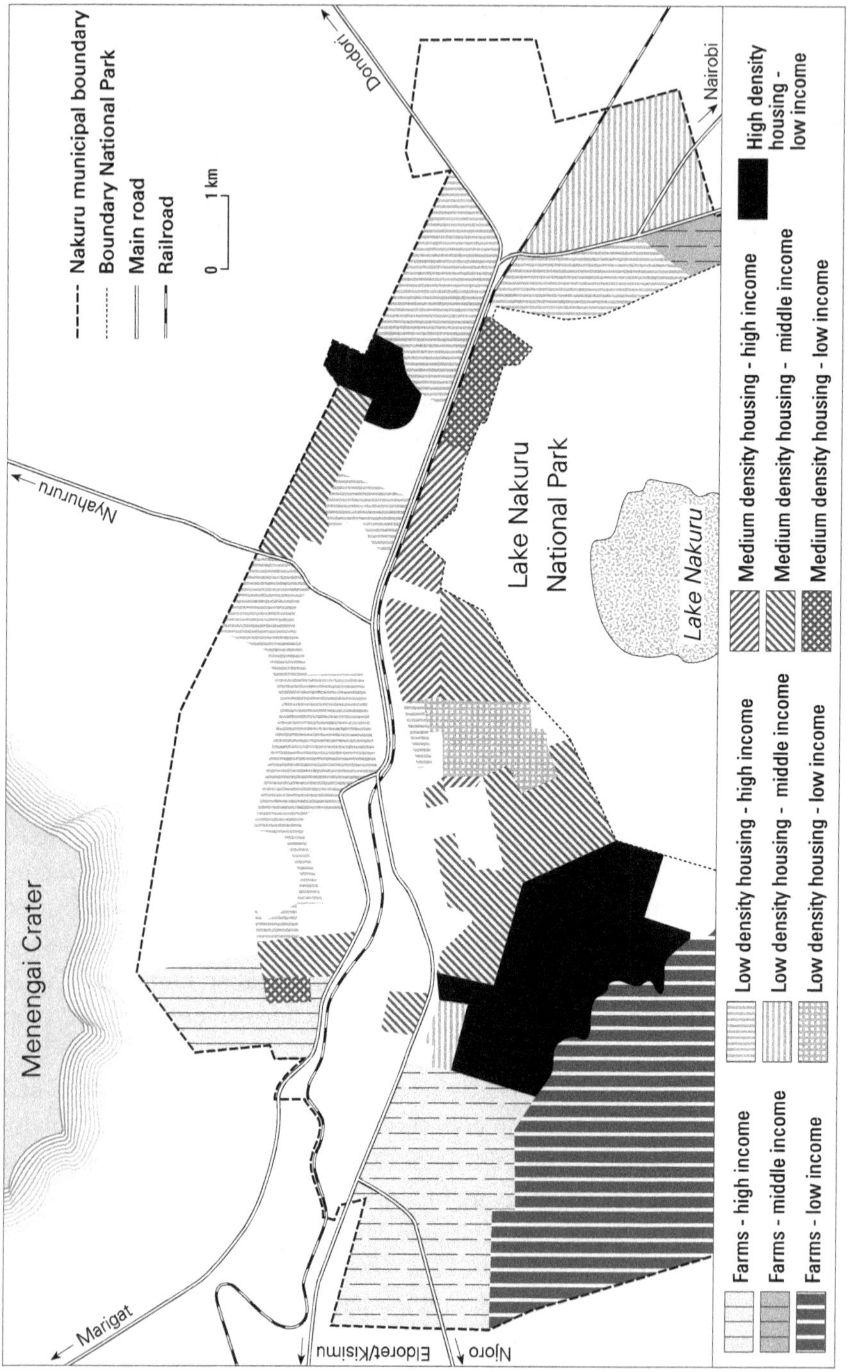

Farming in Nakuru town[5]

By the mid-1990s, farming practices within the town's boundaries were – reluctantly – tolerated by the Municipal Council. However, in cases of complaints or nuisances, the Council did sometimes take measures, legally backed by the Public Health Act. The problem for the municipality was that its enforcement capacity has always been too small. Farming in town has consequently become a common phenomenon and among all categories of the population. Nowadays, the municipality allows crop cultivation as long as the crop is less than one metre high, and although this excludes maize, the crop can be seen everywhere. Most people cultivate common food crops for their own consumption. Crops like kale, cow peas and spinach are also cultivated for commercial purposes, as there is a ready market for these products.

Many people also keep one or more animals. According to information from the local branch of the Ministry of Agriculture, there were about 160,000 chickens in the municipality by the end of the 20th century, 25,000 head of cattle (of which 23,000 under free range and 2,000 in zero-grazing), 3,000 goats, 3,500 sheep and 1,500 pigs. In addition, there were five farmers who kept bees.

According to the Public Health Act, farming is prohibited if it causes a nuisance. Fly breeding, mosquito breeding, disposal of dirty water, pollution of wells, foul smells etc. are all considered nuisances, some of which can bring disease, such as malaria, typhoid, cholera, diarrhoea etc. Pigs usually cause the greatest nuisance.

One of the municipal officers distinguished three types of urban agriculture in Nakuru. First, there is farming in privately owned compounds (on-plot farming). A wide variety of farming activities can be found here because there is little control of such farming by the municipality. Second, there is farming in the compounds of the municipal residential estates. These are rented houses, but farming is very common either in the compounds of the individual tenants or between the housing blocks. Finally, off-plot farming by poor people on land that does not belong to them also occurs in Nakuru. According to an informant, this type of farming is quite common too.

"Localising Agenda 21"

Nakuru is one of the three towns in the world where *Localising Agenda 21: Action Planning for Sustainable Urban Development* is being implemented.[6]

---

5 The information in this section was collected before the 1999 survey and is largely based on personal communication with representatives from the municipality, the district, the Catholic Diocese of Nakuru, several community-based organizations and some urban farmers. What is presented here reflects the perceptions of several key informants on urban agriculture in the town at that time.

6 The others are Essaouira in Morocco and Vinh City in Vietnam (see Tuts 1998).

This programme, launched by UNCHS (Habitat), is a direct result of the Earth Summit held in Rio de Janeiro in 1992 and is based on an awareness that urban development and environmental protection have to be harmonized to make urban development sustainable.

The programme's objective is to provide training to develop a new approach towards urban planning and management, focusing on environmentally aware development of Nakuru ("People's Green City"), with particular attention to low-income groups. The first step was the organization of a Consultation Workshop in 1995 that brought together a wide range of stakeholders in Nakuru, including councillors, officers of the Municipal Council, District and Province, research and training institutions, parastatals, non-governmental organizations (NGOs), community-based organizations (CBOs), industrialists and others (Mwangi 2001). The workshop reached a common understanding of the factors that were promoting and hindering the sustainable development of Nakuru. The result of the workshop was an *Urban Pact*, expressing a vision of desirable development in Nakuru Municipality.

The plan envisaged Nakuru as an 'eco-town',[7] integrating natural and human imperatives. Inevitably, urban agriculture has to be an integral part of this vision because it is a fact of life that cannot be ignored when the sustainable development of a town is being planned. As Kulshreshtha (1998: 47) puts it:

> The assimilation of agriculture as an integral urban function and the protection of the agricultural lands from the threats of expansion of other urban land uses, have become important development imperatives for Nakuru. In the face of slow industrial growth and investment in Nakuru, urban agriculture reveals its potential as a viable, eco-friendly and sustainable development option. This option calls for reorienting and harnessing its potential for urban employment, for economic growth and even more importantly, for living in harmony with nature – a fact that is rooted in the being of Nakuru.

However, during discussions with the municipal officers it was admitted that, although they knew that urban agriculture was everywhere in the town, there was no systematic knowledge of the phenomenon because no relevant study had ever been done in the town.

Other decisions taken during the Consultation Workshop were to set up a town planning unit, prepare the Strategic Structure Plan (see MCN 1999), identify priority zones, strengthen local institutions and stimulate innovative partnerships (Mwangi 2001). A number of local environmental action groups were created to develop environmental action with the municipality. One such

---

7 The other visions being a railroad town, a center of eco-tourism, a regional capital and a service centre, and a prototype town of the East African highlands (Mwangi 2001).

group was the NAROKA group, formed in 1997 and actively involved in solid waste management and water supply initiatives.

During a second workshop, in September 1996, several priority zones were identified requiring immediate planning interventions (Mwangi 2001), for example the further expansion of the town on the eastern side.[8] This rapidly urbanizing peri-urban zone combines residential and agricultural land uses. It was agreed that selective urbanization and the protection of agriculture should receive high priority in the area. Other concrete action plans concerned such topics as water management (water is a scarce commodity in this semi-arid climate), solid waste management, the greening of the town (e.g. by means of tree nurseries in primary schools), the promotion of ecotourism, and awareness training.

The final *Strategic Nakuru Structure Plan* was approved in April 2001. According to Mwangi (2001: 17), "it is the blueprint for urban sustainable development for the town and is probably one of the most important achievements of the LA21 programme". However, the role of urban agriculture is conspicuously absent in this document. The only time the activity is mentioned, farming in town is considered as a temporary feature: "Economically, urban agriculture is a transitory activity which eventually gives way to more traditional urban functions" (MCN 1999: 44). This seems to contradict the remarks by Kulshreshtha's cited above.

So far, a lot has been put on paper, but concrete results have been few. Mwangi (2001) mentions a number of constraints, such as lack of funds, low levels of education on the part of CBO members and County Councillors, frequent personnel changes because of elections and/or transfers, the current economic crisis and the unpredictable political situation. Another problem is the exclusion of certain groups in the whole LA21 process, groups that were supposed to be involved particularly because of their vulnerable position – women, youth and the poor.

The study population

What exactly is meant by urban agriculture in Nakuru Municipality? This question had become very relevant among the municipal authorities in the context of the LA 21 programme, but besides the observation that the phenome-

---

8 Expansion on the western and southern sides is undesirable; in the west because the area is geologically too unstable due to several fault lines, and in the south because of the proximity of Lake Nakuru National Park and the danger of (further) pollution of the lake. In the north, expansion is impossible because of the Menengai Crater.

non is omnipresent in town, specific knowledge was lacking. It was therefore decided to start the research project with a general survey (in 1999) of 600 households thought to be representative of the entire Nakuru population. However, the study was confined to the *built-up area* of the town. The area between the built-up area and the municipal boundary, where farming is still a 'common way of life', was not included in the study. The major objectives of the survey were (1) to collect basic data on farming by the Nakuru townspeople, and (2) to provide the municipal authorities of Nakuru with information on urban agriculture. A second survey of a smaller sample was carried out one year later and focused specifically on food consumption and nutrition.[9]

Some basic characteristics of the sampled households in the two surveys are shown in Table 2.1.[10] The average household size of the 1999 study population was exactly four, which is somewhat higher than the average household size of the whole Nakuru population which was 3.5 in the same year (Kenya 2000). This could be an indication that lower income groups were slightly over-represented in the study population. The average household size of the 2000 study population was much higher (5.7), which was due to the fact that the sample had to consist of households with small children.

Table 2.1 Characteristics of the sampled households

		1999	2000
	(N=)	(594)	(136)
average household size (members)		4.0	5.7
% 'low-income' households*		81.8	55.9
% female-headed households		19.7	10.3
% household heads 30-49 years of age		55.1	64.7

* Monthly cash income less than Ksh. 10,000.
Source: Annex 2, Table A2.3.

The large majority (over 80%) of the 1999 households fell into the category of low-income households (Table 2.1), at least in terms of monthly cash income (<10,000 Kenyan shillings)[11] at the time of the survey and according to the respondents' estimation and willingness to provide the right figure. Over half (53%) of the households could be categorized as 'very poor' (<Ksh. 5,000 a month; see Annex 2, Table A2.3). The better-off households (>Ksh. 20,000 a month) formed a small minority (6%).

9 The sampling procedure for the two surveys is described in Annex 2.
10 For more details, see Annex 2, Table A2.3.
11 The exchange rate at the time was about 70 Kenyan shillings for one US dollar.

One fifth of the 1999 households were headed by a woman (Table 2.1). We could add here that female-headed households were much more common among low-income households than among better-off ones: 69% of female-headed households appeared to be 'very poor', compared with 49% of male-headed households.

The age distribution of the 1999 household heads shows the usual distribution, with over half being between thirty and fifty years of age (Table 2.1). The largest single age group was the one between 30 and 39 years. Very few of the household heads were either younger than twenty or older than sixty. The oldest household head was 77 years of age.

Compared to the 1999 study population, the population in the 2000 survey showed some distinct differences (Tables 2.1 and A2.3). This is firstly because the 1999 study population was a representative selection of the total Nakuru population, while the 2000 study population was *not* because of the focus on the *comparison* between urban farmers and non-farmers; and secondly several selection criterions were applied in 2000 to allow for a worthwhile comparison between the two groups (see Annex 2). As a result, compared to the 1999 study population, the households in the survey in 2000 were on average somewhat larger, somewhat better-off, less often headed by a woman, and on average somewhat older.

Photo 3 Crop cultivator in Lanet uprooting seedlings.
(Sam Owuor, 2000)

Photo 4 Labourer feeding chickens in Rhonda Weavers.
(Sam Owuor, 2000)

3

The farmers

Co-author: Samuel O. Owuor

Who are the farmers in Nakuru town? Two types of farmers can be distinguished, namely those urban residents who farm *in* town and those who farm in the rural areas, denoted as *urban* farmers and *rural* farmers respectively. Two comparisons are thus made in this chapter, the first between urban farmers and those who do not, and the second between rural farmers and those who do not farm in the rural areas. It should be remembered, however, that the focus of this chapter – and indeed of the whole book – is *urban* farmers. But before embarking on the said comparisons, a section dealing with the numbers as well as the spatial distribution of the urban farmers in Nakuru town is presented.

Numbers and geographical distribution

Table 3.1 shows the numbers of households in Nakuru town doing some kind of farming: the farmers of Nakuru town. A distinction has been made between area (i.e. either in town or in the rural area, or both) and the type of farming (either crop cultivation or livestock keeping). The 'total' column reveals that 447 households or 75% of the sampled population were performing farming activities in 1999 in one way or another. Almost all of these households cultivated crops, many of them mixing crop cultivation with livestock keeping. Only a few households (16) kept livestock without growing crops.

Table 3.1 shows that over one third of Nakuru's population could be considered as urban farmers in the strict sense, i.e. they farmed within the municipal

Table 3.1 Numbers of households practising farming in 1999 by area and type of farming (N=594)

	total		urban		rural	
	N	%	N	%	N	%
farming	447	75.3	209	35.2	366	61.6
- crop cultivation	431	72.6	160	26.9	361	60.8
- livestock keeping	299	50.3	121	20.4	222	37.4

Source: 1999 survey.

boundaries. It is, however, possible that this is an underestimation. According to Baba Esther in Kabachia,[1]

> Everybody is practising some urban agriculture. Especially with these retrenchments, you need something to fall back on.

And Mama Barbara, also living in Kabachia, stressed the logic of growing crops if you have space:[2]

> If you have the space then you should use it. The houses in Kabachia come with a big compound. Why should you leave the whole compound idle if you can grow some small things with hardly any trouble? Almost all people around Kabachia are growing something, only you don't see it from the street!

Baba Christopher in Ziwani gave a possible explanation as to why the figure of one third might be too low:[3]

> You know, I think everybody has a small plot somewhere in town. But not everybody will tell you. Some people are afraid that since a lot of those plots are on railway, government or municipal land, they will be forbidden to grow crops.

Forty-two per cent of urban farmers in Nakuru cultivated crops only, 23% kept livestock only, while the remaining 35% did both. Table 3.1 also shows that *rural* farming was more common than farming in town: almost two-thirds of Nakuru households farmed in the rural areas. Nearly all of them cultivated crops or practised mixed farming there and very few kept livestock only.

Based on the percentages under 'urban' in Table 3.1, it is possible to assess the absolute numbers of households in Nakuru town who practised urban farming, those who were cultivating crops and those who were keeping livestock. Table 3.2 shows that the 35% of the sampled population engaged in urban agriculture constitute about 24,000 households. Even so, some 18,400

1 Versleijen 2002: 40. Baba Esther means 'father of Esther'. It is common in Kenya to address people in this way as soon as their first child has been born.

2 Ibid., p. 65. As in the previous footnote, Mama Barbara means 'mother of Barbara'.

3 Ibid., p. 80.

Nakuru households cultivated crops and about 14,000 kept livestock (all in the built-up area).

Table 3.2 Percentages of households farming in town, 1998

	percentage in survey (N=594)	estimated number of households in Nakuru town*
farming in town	35.2	24,000
cultivating crops in town	26.9	18,400
keeping livestock in town	20.4	14,000

* Based on 68,436 households in 1999 (Kenya 2000).

From the 2000 survey, it is possible to assess the dynamics in the numbers of urban farmers for three consecutive years (1998, 1999 and 2000).[4] Overall there appeared to be a fairly sharp decline, i.e. in 2000, the numbers of both crop cultivators and livestock keepers were about 30% lower than in 1998. Interestingly, crop cultivation saw the sharpest decline between 1998 and 1999 (20%), but in livestock keeping this was between 1999 and 2000 (25%). Further analysis shows that the picture was even more complicated. For instance, of the 35 urban farmers who only cultivated crops in 1998, 15 abandoned crop growing (and thereby urban farming), while 8 turned to livestock keeping (either abandoning crop growing or adding livestock to their cropping activities). Of the 42 'mixed farmers' in 1998, some abandoned urban farming altogether, while others dropped livestock keeping and remained with crop growing (the reverse was rare). The group *not* engaged in urban farming was the most stable because the large majority stayed that way; yet, one tried livestock keeping in 1999 and another three started crop cultivation (of which two continued into 2000). The most likely explanation for the overall decline in the number of urban farmers was the drought in 1999 and 2000. As shown in Chapter 2 (Figure 2.1), 1999 was a bad year in terms of rainfall. The following year was even worse, with only 600 mm of rainfall and hardly any rain during what was supposed to be the long rains. Thus, there was little to plant for those who were dependent on rainfall for crop cultivation.[5]

Urban farmers can be found in all parts of the town. However, in some areas they are more common than in others. Figure 3.1 presents the distribution of

---

4 Because of the sampling criteria for the households in the 2000 survey (see Annex 2), it was known whether they were cultivating crops and/or keeping livestock in town during these three years.

5 This was over half of the Nakuru crop cultivators (see Table 4.2 in Chapter 4).

farmers in town by housing density.[6] In general, there were more urban farmers when housing density was lower. Of the eight respondents in the only very-low-density area of Milimani, seven were urban farmers and all except one were both crop cultivators and livestock keepers. Medium-density areas like Kabachia and London/Menengai also had high percentages of urban farmers, 82% and 71% respectively. In the high-density areas of Rhonda Muslim and Rhonda Kaptembwa, on the other hand, only about 15% of the households were engaged in some kind of agriculture in town. However, this reverse relationship between housing density and numbers of urban farmers applied particularly to crop cultivation. Livestock were more evenly spread over the estates, not only small livestock but also larger animals. Keeping one or a few animals requires relatively little space, particularly if they are kept in zero-grazing or if the animals roam freely in the streets.

Figure 3.1 Farmers in town by housing density (%)

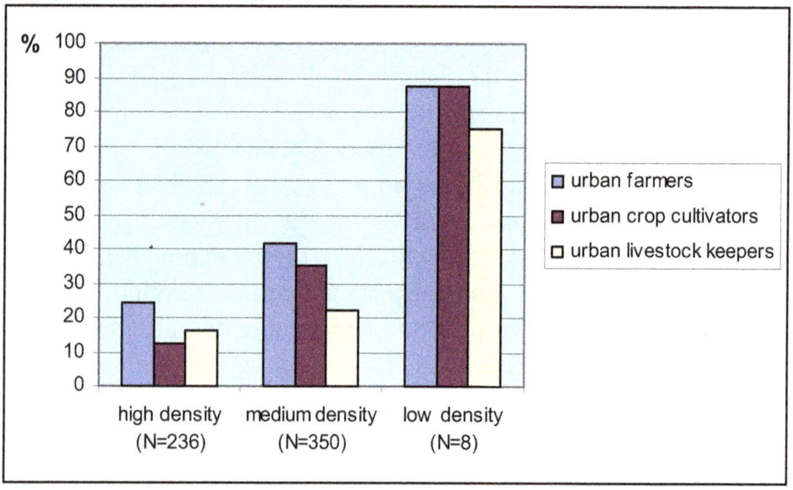

Source: 1999 survey.

Urban farmers and non-farmers

For the purpose of comparison, urban crop cultivators and urban livestock keepers have been combined to one group of 209 'urban farmers'. The group of urban 'non-farmers' consists of 385 households.

---

6 The percentages by research cluster are presented in Annex 3, Table A3.1.

Household characteristics

Table 3.3 presents a summary of some of the major household characteristics of the two groups.[7] There is a difference in household size, with farming households being larger. Another important distinction between the two categories concerns household income. Although one always has to consider income data with great care, it is clear that the poorest households – i.e. with a monthly income of less than Ksh. 5,000 – were under-represented among the urban farmers. This is confirmed by the variable on house ownership: households owning the house in which they lived were quite over-represented among the farmers (although the category of households renting their house was by far the largest in both groups; see Table A3.2). The fourth variable presented in Table 3.3 – the housing density of the estate in which the household is located – is more of a geographical variable than a household variable. Farmers were somewhat over-represented in the less densely housed areas (Table A3.2), non-farmers in the high-density estates. This is in line with the figures presented in Figure 3.1 and is undoubtedly related to the availability of space.

Table 3.3 Urban farmers and non-farmers: summary of household characteristics (%)

		urban farmers (N=209)	non-farmers (N=385)
household size (no. of members)	5 or more	57.4	22.9
household income (Ksh/month)	up to 5,000	33.2	64.0
	more than 10,000	30.8	11.4
house ownership	own house	21.5	4.9
housing density of estate	high	27.3	46.5

Source: Annex 3, Table A3.2.

Characteristics of household heads

Farmers and non-farmers were compared on a number of characteristics of the household heads.[8] On the whole, there were no major differences between the heads of the farmers' households and those of the non-farmers' households. In both groups, the majority were male, were permanently resident in the household, were married, had at least secondary school, and had either a steady job or

[7] See Annex 3, Table A3.2 for details.

[8] Age, sex, ethnic group, type of residence, marital status, educational level and occupational status. Details are presented in Annex 3, Table A3.3.

were self-employed. There was one important difference, however: the heads of farming households were generally older than those of non-farmers. This can be related to household size given in Table 3.3. Urban farmers generally have larger households, i.e. more mouths to feed. Households in the early stages of the 'family life cycle' – young and small – are clearly under-represented among the farmers.

The household heads were also compared regarding their migration histories.[9] In both groups, the large majority had not been born in Nakuru town. They came from all over Kenya, but particularly from the central and western parts. On average, the farmers had come to Nakuru before the non-farmers. Of the latter, the majority had arrived in Nakuru over the last ten years, i.e. twice as many as among the farmers. This difference should be seen in relation to the age of the household head: the farmers were on average older. Finally, the reasons for coming to Nakuru did not differ between the two groups. The large majority came in order to work in Nakuru or to look for work.

Urban non-farmers

The non-farmers were asked why they had not cultivated crops or kept livestock in town in 1998. A summary of the reasons mentioned is presented in Table 3.4.[10] It is clear that although a wide variety of reasons were given, the land issue was the dominant one, followed by a lack of other resources, legal considerations and various other reasons.

The lack of access to land within the municipality is by far the most important reason for non-farmers not being engaged in agricultural activity (Table 3.4). This applies more to crop cultivation than to livestock keeping because more land is required for growing crops than for keeping (small) animals. Some households do indeed keep their animals inside their houses. For some non-farmers there was no need to farm in town because they had access to a plot in the rural areas. However, for many others this is not a reason for not farming in town as well (see below).

Lack of land is related to the town's expansion. For instance, one respondent said that her husband had acquired a piece of "idle open land" of about half an acre in 1975. For twelve years, she cultivated maize and beans there, which was enough to feed her household for about six months a year. However, in 1987, the Municipal Council of Nakuru repossessed the land for expansion purposes and the only plot left to them was the small *shamba* bordering their house. Another respondent recalled that between 1963 and 1978 she sold vegetables she had personally cultivated on open spaces not far from her estate.

---

9 For details, see Annex 3, Table A3.4.
10 For details, see Annex 3, Table A3.5.

> Nakuru was not developed as it is now. The only developments in most of these areas were the Nakuru Municipal Council housing estates like Kaloleni, Abong' Lo Weya, Flamingo and others. Open spaces, which we used for farming, were many and nobody bothered with us. Many of these open spaces were undeveloped Municipal Council land. I had three plots not far from each other. They were not very big. I think less than half an acre each. It was not advisable to take a big plot because of security and fear of losing the plot when the owner reclaims it. Three or four different people could cultivate a plot of about one acre. Of course they had other smaller plots elsewhere.

She added that there was no rent to be paid because the owners were "more than happy that somebody was taking care of their plots".

Table 3.4 Non-farmers: reasons for not farming in town by type of farming (%)

		no crop cultivation (N=434)		no livestock keeping (N=473)	
		reasons (>100%)	main reason	reasons (>100%)	main reason
land issues	no access to urban land	85.7	75.6	74.4	62.8
	have access to rural land	8.1	2.3	5.7	1.9
lack of other resources	no capital	28.6	9.4	24.1	10.6
	lack of time	7.1	4.4	6.6	4.4
legal considerations	harassment	1.8	0.2	4.7	1.7
	myself/landlord disapproves	5.3	0.9	11.4	3.8
other reasons	not worthwhile	5.3	2.5	7.2	5.1
	had not thought of it	0.9	0.7	2.5	2.5
	was not in Nakuru	2.3	2.1	1.7	1.7

Source: Annex 3, Table A3.5.

As a result of the increasing scarcity of farming land in town, it has become increasingly expensive to rent land. Another respondent compared the situation before 1990 with the present situation as follows:

> It was very cheap to rent a plot in the municipality by then. With not more than five shilling, you could get a sizeable plot to rent on a monthly basis. For those who liked farming, renting a plot was a normal and common thing to do those days. Nowadays, to rent a plot within the municipality, that is if you are lucky, costs not less than 6,000 shilling per year for an acre.

These developments could explain why lack of capital was the second most important reason not to farm in town (Table 3.4). As one respondent said, "due to lack of access to urban land and capital to rent a plot, I never engaged in urban crop cultivation". And yet another respondent explained that "I was forced to stop growing crops by 1993 because of not only a lack of plots but also the high rents imposed on those available". For some, however, lack of *other* resources, such as time and/or labour, was a reason not to farm in town. Lack of labour can be due to ill health, like the 56-year old woman who had to stop growing crops because of poor health after an operation.

Legal considerations were more often mentioned as a reason not to keep animals than not to grow crops (Table 3.4). This is probably due to the fact that livestock keeping is generally considered more of a nuisance than crop cultivation. As Tables 3.4 and A3.5 show, the disapproval of farming in town can come from various levels: the municipality, landlords, neighbours and, finally, the people themselves. Harassment was not an important reason not to farm in town, although several respondents mentioned this in passing. For instance, as one recounted, "harassment by the local authority by way of slashing the crops every now and then was a constant threat", especially for the off-plot cultivators. And according to the same person, even her ducks were not safe:

> I stopped with keeping ducks in 1996 when harassment intensified. In that year, the local authority contaminated all the open drains in the estates with poison and when the ducks fed on the dirt from the drainage, they all died.

And the respondent who refrained from farming because of lack of land and lack of capital to rent a plot added that she did not want to be harassed by the Municipal Council who "keep on destroying other people's crops".

The other reasons include people for whom farming in town was simply not an option (Table 3.4). Particularly among non-livestock keepers, for 10% of them it was either not seen as an activity which was worthwhile or the respondent had never considered it. Finally, for a few people, the question was not applicable because they did not live in Nakuru town in 1998.

Some of those households *not* practising urban farming in 1998 had actually done so before: 34 had cultivated crops and 35 had kept livestock. A variety of reasons were given why they had stopped farming, for instance theft of the crops, the plot being used for another purpose, the plot having been repossessed by the owner, a lack of rain, and insufficient profit. The reasons for stopping with keeping livestock were much less diverse and focused on the problems of pests and diseases, theft and insecurity.

Rural farmers and non-farmers

In what ways do those who farm in the rural area (rural farmers) differ from those who do not (non-farmers)? Table 3.5 summarises some household characteristics of the two groups.[11] The three household characteristics in Table 3.5 point to three possible reasons for urban households practising rural farming: the number of mouths to fill (household size), the purchasing power of the household (income class), and the amount of space in the (urban) residential area (housing density of the estate). The figures show that none of these variables seem to determine whether or not people engage in rural farming. As shown in Table 3.3, household size does seem to be an important determinant in practising *urban* farming, but it is not so for *rural* farming. Poorer households do not practise rural farming more often than richer households. And lack of space for *urban* farming in the urban residential area seems not to be compensated for by a higher frequency of *rural* farming.

Table 3.5 Rural farmers and non-farmers: summary of household characteristics (%)

		rural farmers (N=366)	non-farmers (N=228)
household size (no. of members)	5 or more	35.2	34.6
household income class (Ksh./month)	up to 5,000	47.8	61.8
	more than 10,000	20.1	15.1
housing density of estate	high	45.6	40.4

Source: Annex 3, Table A3.7.

The question as to whether rural farming and urban farming are substitutes of each other can also be reworded as follows: Do urban dwellers who practise *rural* farming refrain from *urban* farming and vice versa? This appeared not to be the case since the percentages of *urban* farmers among both rural farmers and non-farmers appeared to be exactly the same, namely 35%. This applies to crop cultivators (25% and 30% respectively) as well as to livestock keepers (20% and 21% respectively). In other words, those urban dwellers in Nakuru who do *not* have access to a *rural* plot are *no* more inclined to engage in *urban* farming than those who *do* have access to a rural plot.

The household heads of the 'rural farmers' and 'non-farmers' showed no big differences.[12] The groups appeared to be very similar in terms of age, type of residence, educational level and occupational status. However, regarding sex, marital status and ethnic group, the situation is somewhat different. The per-

[11] For details, see Annex 3, Table A3.6.
[12] For details, see Annex 3, Table A3.7.

centage of female-headed households among non-farmers was twice as high as among farmers, while the same applied to the percentage of single/divorced/-separated/widowed household heads. These two variables are to some extent related because many of the 'one-adult households' (single, divorced, separated or widowed) were female-headed households and one could imagine that for these households it would be much more problematic to practise farming outside town than for households with a head and a spouse.

As far as ethnic background is concerned, it is conspicuous that the Kikuyu (accounting for 44% of all household heads and by far the largest group in Nakuru town) practised much less rural farming than the other major ethnic groups. Of all the Kikuyu households in Nakuru, about half were farming in the rural areas compared with about two-thirds to over three-quarters of the Luo, Luhya, Kalenjin, Kisii and Kamba.

In sum, rural farmers and non-farmers show no major differences. Hence, farming in rural areas seems to be more a matter of opportunity – i.e. whether one happens to have access to a plot – than of necessity. For most urbanites, access to a rural plot is first of all seen as a fall-back mechanism. This was described by Baba Christopher as follows:[13]

> At a certain moment, most people return to their rural home when they retire and go farming again. I will go back there to practice farming and my children will go with me. Our rural home is the place we came from and where we have our origins. You know, we will never forget that we are farmers and that we can always fall back on farming. The only problem is the land, like my father had to buy more land for his wives and children, so do I have to buy more land for my children and my sons will have to buy more land for their children. Because they will all be farmers, whatever other occupation they will have. Growing crops, no matter how much, also gives us some security. You know, your boss can delay your payments for half a year but then at least you can eat from your own *shamba*.

And another respondent phrased it thus:[14]

> Having a homestead at home is very essential, especially for us Luos. It plays a very important role because I can be retrenched from work at any time. That is the place I will go back to. I cannot afford to stay in Nakuru. When anything happens to me or my wife or my children, then we have a home to be buried instead of being thrown in the public cemetery here in Nakuru. That is our home and never shall we stay in Nakuru forever because these are just but rental houses.

In summary, this chapter has shown that the majority of the population of Nakuru town were engaged in farming activities. Farming in the *rural* areas (usually at the rural 'home' at some distance from Nakuru) appeared to be more

---

[13] Versleijen 2002: 33

[14] Owuor, field data, 2003.

common than farming in town. Nevertheless, the latter could be found all over town, although more so where housing density was lower. Not surprisingly therefore, the main reason for most of those *not* farming in town was a lack of access to space or insufficient money to rent a plot. Urban farming households were generally larger (more mouths to feed) than non-farming households. The low-income households appeared to be under-represented among urban farmers but because the large majority of the Nakuru population belong to this income category, in absolute terms the poor were still the largest group among the urban farmers.

Photo 5 Crop cultivation in Kabachia (maize, bananas).
(Sam Owuor, 1999)

Photo 6 Ziwani estate.
(Dick Foeken, 1999)

4

The crops

Co-author: Samuel O. Owuor

Urban plots

Table 4.1 presents information about the various characteristics of the urban *shambas*.[1] Most of the 180 plots used for crop cultivation were located in the farmers' own compounds (61%). This is the category of 'on-plot' or 'backyard' farmers. The rest of the plots were located elsewhere ('off-plot' or 'open space' farming): in the respondent's estate (17%), along a railway (8%), in another estate (6%), along a road (4%), or in another location (5%) such as along a river, under a power line, next to a cemetery, a park or a sewer or in a school compound. The percentage of plots at one's own compound is somewhat higher in lower-density estates, Milimani having the highest percentage (100%). Nevertheless, over half of the plots in the high-density areas were also in the people's compounds.

The ten plots of the households in medium-density Ziwani were all located outside the estate itself, because the landlord (the railway company) does not allow farming in the estate, even though there is space to do so (see Photo 6). The plots are located along the railway line to Nairobi, a zone to which the residents of this railway estate have easier access than others. Because the estate lies at some distance south of the railway itself, Ziwani is also the only area where the majority of the plots were located at a distance of more than half-an-hour's walk from the house. In most other areas, the large majority of the plots

---

[1] For details, see Annex 4, Table A4.1.

were within a ten-minutes walk (Table 4.1), except for Flamingo I (medium density), where a third of the plots were located at a distance of more than an hour's walk. This has to do with the back-to-back construction of the houses and the resulting absence of backyards.

Table 4.1 Summary of characteristics of urban plots by housing density (%)

	(N=)	high density (35)	medium density (138)	low density (7)	total (180)
location	on own compound	52.9	60.4	100.0	60.6
	within estate	29.4	14.2	-.-	16.6
distance to plot	<10-minute walk	65.7	69.6	100.0	70.0
plot size (m^2)	<100	22.9	44.9	16.7	34.1
	1000+	37.1	28.3	83.3	31.8
ownership of plot (%)	own land	57.1	25.5	57.1	33.0
	landlord	28.6	50.4	42.9	45.8
	government	2.9	17.5	-.-	14.0

Source: Annex 4, Table A4.1.

As mentioned in the previous chapter, average plot size was 964 m^2. The smallest plots were a few of just one square metre. The largest one was a plot in Lanet that measured 16,000 m^2 (1.6 hectares). There is no clear relationship between plot sizes on the one hand and housing density on the other, although it is not surprising that six of the seven plots in Milimani measure more than 1,000 m^2 (Table 4.1). The relatively high percentage of small plots (less than 100 m^2) in the medium density areas may be explained by the fact that, compared with the households in the high-density areas, more plots there are located within the households' own compounds.

Plot size can to some extent be determined by input factors. This is exemplified by the case of Baba Christopher (as described by Versleijen 2002: 36) who cultivated a plot of 50 m^2:

> Baba Christopher would not want a bigger plot than he has now because of, first, the availability of labour, second, the needs of the family and, third, the amount of seeds and seedlings they can afford to buy. By cultivating a plot of 50 m^2, he is able to feed his family from the *shamba* in such a way and for such a period that he can feed them from his salary for the rest of the year and even educate them and meet other expenses such as hospital bills. To cultivate a larger plot would mean that they have to buy extra seeds. Right now, all the seeds they use are from last year's harvest, so they do not incur any expenses in the planting season.

One third of the plots used for crop cultivation were owned by the cultivators themselves (Table 4.1). Another 46% of the plots were owned by a landlord, while 14% of the plots were on government land. It is conspicuous that the percentage of 'own land' in the high-density areas is much higher than in the medium-density areas and that it is the other way around with plots owned by a landlord or by the government. Nearly all residential land in the built-up area of Nakuru town is government land that is leased out to the residents. It seems that the inhabitants of the lower-density estates (with usually higher levels of education) are more aware that they are not the actual owners of the land than those living in the higher-density estates. The large majority of the plots are compounds of rented houses, which is confirmed by the fact that hardly anybody pays rent for the plot.

Crops and inputs

Choice of crops

A wide range of crops was cultivated in Nakuru in 1998. Table A4.2 (Annex 4) offers a full list of all the crops cultivated on the 180 plots. Both from Annex 4 and from looking around the fields it is clear that mixed and inter-cropping were common, though the majority of the plots (58%) had no more than three crops. In eight cases, ten or more different crops were found on one single plot. On about a dozen plots, only one crop was being cultivated.

The average number of crops cultivated per household in 1998 was 4.3. A small minority (8%) cultivated just one crop. About two-thirds (64%) of the crop cultivators had planted two to five crops, and another quarter (26%) between six and ten crops. There is no relationship between plot size and the number of crops per plot. For instance, on about two-thirds of both the smallest plots (<10m^2) and the largest plots (1,000+m^2), one to three crops were being cultivated. The largest variety of crops was found on plots measuring between 100 and 1,000 m^2. On his plot of 30 m^2, Baba David cultivated four crops, namely maize, beans, potatoes and *sukuma wiki* (Versleijen 2002: 43).

The three crops that stood out as by far the most important in terms of the number of households cultivating them were kale (*sukuma wiki*), maize and beans. Kale and maize were grown by about two-thirds of the crop cultivators, and beans by almost 60%. Onions, spinach, tomatoes and Irish potatoes were cultivated by 20% to 30% of the cultivators and cowpeas, bananas and spider plant (*saget*) by 10% to 20% (for exact figures, see Table A4.2).

Sukuma wiki is the local name for a green, leafy vegetable of the spinach variety (*Spinacea oleracea*) and also called kale, literally meaning "to push the week". This refers to the importance of the crop for subsistence dwellers in their daily diet due to its high yield and low price. People with low incomes can survive on it, especially during the last week of the month ("push the week") before salaries are paid. It is a fast-growing crop, especially in the red soil areas in the town, and has a high nutritional value: its high calcium and phosphor contents are almost comparable with that of whole milk (Sehmi 1993). For these reasons, and because it is relatively cheap, *sukuma wiki* is a typical ingredient in the diet of poor households and favoured as the usual supplement to the basic *ugali* dish (stiff maize porridge). The importance of the crop is illustrated by respondents' statements quoted in Versleijen (2002):

> When you grow kale you at least know you can eat. (p. 65)

> We cultivated *sukuma wiki* on the whole compound and sold large amounts of it. Although we would have got more money out of the chickens, the benefit of the *sukuma* was that it was a low investment. (p. 63)

> *Sukuma wiki* you can easily grow. It is cheap to grow, it hardly needs any care and it is resistant to drought; although of course it would not survive a real drought, but it can sustain longer than other crops. Almost everybody is growing *sukuma*, except maybe just a few, but I think that it is hard to find someone around here [Kabachia; DF] who is not growing some *sukuma*. (p. 79)

> Actually, to keep *sukuma* is not really a decision, you just do it, like everybody does it. (p. 70)

To some extent, the type of crops cultivated depends on the location of the plot. The variety of crops cultivated in the homesteads was much larger than on plots located elsewhere. Typical compound crops were kale and bananas and to a lesser extent spinach, onions and tomatoes. Kale was grown on 80% of compound plots and on 32% of the other plots. Bananas were almost exclusively found in compounds. Maize and beans were found on about half of compound plots and on 75% and 68%, respectively, of plots elsewhere. Versleijen (2002: 62) describes the compound of Mama Esther:

> The area Mama Esther cultivated is split up into three pieces. The biggest piece is about 15 m^2, on which she grows *sukuma wiki*, *kunde*,[2] *saget*[3] and different types of *mchicha*[4]. On the second one, which is about 4 m^2, pumpkins are grown. The pumpkins hardly bear any fruit but the main use of the plant is the stamped leaves. On the third piece, of about 10 m^2, there is *mchicha* as well. If one looks carefully, some tomatoes can be found here. (…) A small part is grown with rosemary and the fence to the neighbour's compound is covered with passion fruit.

It is remarkable that even on the smallest plots ($<10m^2$) all ten of the major crops were represented. For instance, maize was found on more than half of

---

[2] Peas.
[3] Spider plant.
[4] Wild spinach.

these tiny *shambas*. On the larger plots (1,000+m^2), maize and beans were more common (77% and 72% respectively), but most other crops were less extensively cultivated. Kale, for instance, was found on only 32% of these larger plots due to the fact that these plots are often somewhat further away from the house: over half of the plots of 1,000 m^2 or more were at least half-an-hour's walk away. On the 14 plots even further away (at least an hour's walk), maize, beans and Irish potatoes were over-represented while the other crops were either under-represented (kale and cowpeas) or not found at all. In other words, distance is a limiting factor regarding the choice of which crops to grow. This is related to the perishability of the crop, the risk of theft and the use of inputs, including labour.

Land ownership is another limiting factor. All crops could be found on plots owned by either the cultivator or by a landlord because over 70% of these plots were located in people's own compounds. Growing crops on government land or on land where the user does not know who it belongs to is much riskier. Hence, mainly maize and beans and to a lesser extent kale and cowpeas could be found on these plots.

The choice of what to grow is to some extent determined by the person responsible for cultivation. Men were more inclined to grow staples like maize and beans than women,[5] while women more often cultivated vegetables such as spinach, onions and *saget*.[6] In other words, women are more inclined to grow a wider variety of crops. This is related to their traditional function as the household's food provider, and their attempts to achieve a more balanced diet.

Inputs for crop cultivation

Inputs in crop cultivation include such factors as labour, tools, material inputs and information. In this section, the first three types of inputs will be discussed, with an emphasis on material inputs. 'Information' is related to 'support' and is dealt with in Chapter 7.

As far as *labour* is concerned, the 1999 survey did not contain a detailed assessment of the exact amounts of labour input. However, we do know who was responsible for the household's crop cultivation. In the large majority of crop-cultivating households, it was either the male head (on 27% of the plots) or the spouse (49%) or the female head (13%). In the remaining cases it was another household member (10%) or a paid labourer (2%). In most cases, the one responsible is not the only household member working on the plot. For

5 Maize was cultivated by 73% of the male heads and 56% of the female heads and spouses. The figures for beans were 71% and 49% respectively. See Annex 4, Table A4.3.

6 Spinach: 10% of the men, 23% of the women; onions: 10% of the men, 30% of the women; *saget*: 4% of the men, 12% of the women. See Annex 4, Table A4.3.

instance, Charles (living in Ziwani) was the one responsible but his wife and children worked on the *shamba* as well. Ann (female head, Kabachia) was helped by her son. Monica (spouse, Lakeview) did it all on her own, on average spending, according to her, about two hours per week on her plot of about 450 m^2. At peak periods, such as planting and weeding, the *shamba* requires full-time work, however. Weeding was normally done once or twice during the growing season, except for kale because these "require constant weeding", as Monica remarked.

Crop cultivation was not usually a full-time job, though 14% of the crop cultivators interviewed said it was. More than a quarter of the crop cultivators had also used hired labour. This appeared to be more common on plots that had been in use for longer and on plots owned by the cultivator him/herself. Richer and bigger households were more inclined to hire additional labour, while female heads rarely did so. If labour is hired, it is mostly for a few days only. Charles hired somebody for just one day to help with the planting. Reuben (Rhonda Kaptembwa) did the same, but for weeding only. James (Rhonda Weavers) hired somebody for both activities, a week in all. Rachel (Mwariki) used hired labour for planting, weeding and harvesting on her half-acre plot, paying Ksh. 100/day for planting and harvesting and Ksh. 150/day for weeding.[7]

As for tools, most of the plots were simply too small to allow any machinery to be used. The tools used by nearly everybody were the hand-hoe and the cutlass. Only in exceptional cases was a tractor used.

Table 4.2 shows the percentages of crop-cultivating households using certain material inputs during the 1998 growing season. Ten respondents said they used no inputs at all. Almost all crop cultivators used at least one type of fertilizer. Most fertilizer was of the traditional (i.e. organic) type: manure, crop residues, urban waste and (in two cases) ash. The manure came either from people's own farms (mixed farming in town) or from a neighbour who kept livestock. Crop residues almost always came from the farmer's own (urban) farm. Chemical fertilizers were used by about one third of the crop cultivators. The use of (chemical) pesticides and insecticides was not widespread – about 30% and 10% of the cultivators respectively. Most farmers used local (traditional) seeds and seedlings, although more than half used improved materials as well. Finally, irrigation was practised by almost half of the cultivators. All except two

7 Except for Rachel, none of the other crop cultivators hired anybody for harvesting. This could largely be explained by the fact that these examples are from the in-depth interviews that were held in 2000 covering the year 1999, i.e. the year when rainfall was insufficient (see Figure 2.1) to allow for a reasonable harvest, if there was a harvest at all. Only those able to irrigate their crops, like Rachel, were able to harvest something.

obtained their water from a tap (even though the use of domestic water for irrigation purposes is illegal). One of the latter two used sewage water for irrigation and the other applied cattle urine.

Table 4.2 Material inputs for crop cultivation in town (%; N=160 households)

type of input	%	type of input	%
no inputs	6.3	chemical pesticides	29.4
		chemical insecticides	8.8
chemical fertilizer	35.6		
manure as fertilizer	53.1	local seeds/seedlings	70.6
crop residue as fertilizer	35.0	improved seeds/seedlings	57.5
urban waste as fertilizer	3.1		
ash as fertilizer	1.3	irrigation	44.4

Source: 1999 survey.

Many farmers tried to economize on the use of inputs, which can be seen in the example of Baba David, as described by Versleijen (2002: 44-45):

> In the case of *sukuma wiki* he does not need to buy seeds or anything because he just removes the smaller plants from in-between the larger plants and plants them on open space. Also suckers can be used. As for maize, he buys seeds in case there are not enough useful seeds from last year's harvest. For beans, it is the same, although those are usually all from last year's harvest.
> He does not always use fertilizers, he only buys fertilizers when he has some extra money. Crop leftovers are left on the plot to serve as fertilizers. He does not use any water (...) because there is no water source around.

Others, however, normally buy their inputs, such as Baba and Mama Joshua (Versleijen 2002: 52):

> The seeds they use for planting for the beans are bought at the market and for maize Baba Joshua buys hybrids from the shops just within the town. For *sukuma wiki* seedlings are bought at the market. It is easier to buy seedlings since they lack the space for a nursery. And to keep a nursery at the plot is not safe because of theft, the more so as everybody has free access to the plot [which was located along the railway; DF]. Furthermore, a nursery needs more frequent and intensive care than they can visit their plot.

Inputs can be classified in various ways. An initial division is between traditional and modern inputs. Manure, crop residues and local seeds can be considered as traditional inputs, while the three chemicals plus improved seeds can be seen as modern inputs. The question behind this classification is whether the use of modern inputs leads to higher yields. A second classification is between chemical inputs and non-chemical inputs, which is important for environmental reasons. In this context, one can also distinguish so-called sustainable (or environmentally friendly) inputs that are organic, can be recycled and may lead to reasonable yields: manure, crop residues and improved seeds. Finally, irrigation is an input that stands on its own: without sufficient water no cultivation is possible. The relationship between these categories of inputs on the one hand and crop yields on the other is dealt with in the next section.

Table 4.3 shows that most crop cultivators (86%) used at least one traditional input, while over half used two or three. Modern inputs were less frequently used and almost 30% of the cultivators used no modern inputs at all and another third only one. Chemical inputs were used even less. Over half of all farmers did not use these types of inputs, while only eight used all three types. The picture of sustainable inputs resembles that of traditional inputs, as two of the three types of inputs categorized under sustainable inputs are traditional. The average number of inputs used in each category (Table 4.3, right-hand column) confirms the general picture that crop cultivation in Nakuru town is quite traditional in nature.

Table 4.3 Number of inputs by input category (%; N=160 households)

category	number of inputs: 0	1	2	3	4	total	average number of inputs
traditional inputs	13.8	35.0	30.0	21.3		100	1.6
modern inputs	29.4	32.5	20.6	12.5	5.0	100	1.3
chemical inputs	51.9	27.5	15.6	5.0		100	0.7
sustainable inputs	16.3	37.5	30.0	16.3		100	1.5
irrigation	55.6	44.4				100	0.4

Source: 1999 survey.

It is hypothesized that the use of inputs varies with certain plot characteristics (size, location, distance, ownership), household characteristics (income, size) and the characteristics of the person responsible for crop cultivation (sex, marital status, educational level, age). To simplify matters, three mutually exclusive categories of material inputs are compared: chemical inputs, sustainable inputs and irrigation.

Since location, plot distance and size were interrelated, it is not surprising that these characteristics showed the same tendencies as far as the use of inputs was concerned. Chemical inputs were used more on plots located *outside* people's compounds, somewhat further away and relatively larger in size. However, for the use of sustainable inputs, these characteristics showed no differences. Sustainable inputs were more commonly used on plots owned by the users themselves and had been in use longer. Irrigation was more often practised in compounds, which is not surprising as most of the water came from people's own taps.

Except for irrigation, there appeared to be very little difference between lower-income and higher-income households regarding the use of certain inputs.[8] As for household size, larger households used more chemical and sustainable inputs than smaller ones. This could be expected since there are more mouths to feed. Irrigation showed no relationship with household size, its use being more dependent on the presence of a (functioning) tap than on anything else.

One of the clearest differences regarding the use of inputs concerned the sex of the person responsible. Men were more inclined to use chemical inputs than women, though women irrigated their plots more often.[9] A further differentiation of the women into spouses and female heads shows that it was particularly the latter category that practised a relatively input-poor type of crop cultivation. Whatever category of inputs is considered, female heads used it less frequently. However, this applies in particular to chemical inputs: only 10% of female heads used any chemical input as opposed to 63% of male heads and 50% of their spouses. This is likely to be due to the usually (very) low welfare level of female-headed households. Other characteristics of the person responsible, such as educational level, occupational status and age, showed few differences in the use of inputs.

Yields

Crop yields are determined by various factors but by far the most important one is the weather – in particular the amount of rainfall and its distribution over the growing season(s). For instance, harvests were quite bad in 1999 and 2000 because of drought. Most of the data presented here concern the harvests of 1998, which was a fairly normal year in terms of rainfall (see Figure 2.1). Other determinants include such factors as labour inputs, material inputs as defined above, farming techniques, etc. The data available are mainly on the types of inputs used, whether additional labour had been hired and any assistance had been received.

---

8 See also Chapter 10.

9 Chemical fertilizers were used by 57% of men and 27% of women. The figures for chemical pesticides were 48% and 25% respectively. Irrigation was practised by 51% of women, compared with 28% of men. See Annex 4, Table A4.3.

Table 4.4 shows the average amounts harvested per crop-cultivating household and per crop type (i.e. the 10 most commonly cultivated crops). At first sight, the harvests of the various crops (column 3, in kg) appear modest. Nevertheless, when looking at, for instance, maize, the 101 households cultivating this crop harvested about 22,600 kg in 1998, which amounts to some 2.6 million kg for the built-up area of Nakuru as a whole.[10] Likewise, Nakuru crop cultivators produced about 1 million kg of kale, 0.8 million kg of beans, 135 tons of onions, 380 tons of spinach, 60 tons of tomatoes, 325 tons of Irish potatoes, 216 tons of cowpeas, 12 tons of bananas and 72 tons of spider plant. If the other 30 less important crops (Table A4.2) were also included, it is estimated that total crop production in Nakuru town in a normal year would amount to some six million kg. And if the peri-urban areas of Nakuru town were to be included, this figure would be much higher.

Table 4.4 Harvests of major crops cultivated in town

crop type	(1) % households cultivating (N=160)	(2) N	(3) aver. amount harvested (in kg)*
kale (*sukuma wiki*)	68.1	109	84
maize	63.1	101	224
beans	58.8	94	75
onions	28.1	45	26
spinach	22.5	36	92
tomatoes	21.9	35	15
Irish potatoes	20.0	32	88
cowpeas	17.5	28	67
bananas	16.9	27	4
spider plant (*saget*)	11.9	19	33

* Only households cultivating that crop (see column 2). During the survey, harvests were given in many different units. To make the figures easily comparable, all units have been converted into kg. As this method implies an element of speculation, the presented average harvests have to be considered as indications only.
Source: 1999 survey.

The average harvest of all crops was almost 300 kg per plot (Table 4.5), which means that with an average plot size of almost 1,000 m^2, productivity, i.e.

10 This was calculated as follows. There were some 18,400 crop cultivators in the town (see Table 3.2), of whom 63.1% cultivated maize, making 11,610 maize cultivators. The average maize yield was 224 kg (Table 4.4), making a total of 11,610 times 224 = 2.6 million kg. The same procedure was followed for the other crops in Table 4.4.

the average amount harvested per m^2, was a modest 0.3 kg.[11] However, there are important differences between the various plot size classes. In general, productivity was higher if plots were smaller. Of course, one could expect a relatively higher output from smaller plots, as many were located in people's own compounds where one is inclined to pay more attention to them than to crops on plots further away.[12] But since many of the plots of the other size categories were also located in the households' compounds,[13] the figures in Table 4.5 suggest that plot size is an important determinant of crop yield.

Table 4.5 Mean harvest (all crops) by plot size

plot size (m^2)		N (plots)	mean harvest (kg)	mean plot size (m^2)	harvest per m^2 (kg)
all plots		168	301	964	0.31
plot size (m^2):	<10	18	121	6	20.17
	10-99	47	175	37	4.73
	100-999	50	191	373	0.51
	1000+	53	578	2670	0.22

Source: 1999 survey.

One explanation for the higher output as plots decrease in size could be the use of material inputs. However, on the whole, the different types of inputs mentioned in the previous section were not used more often on smaller plots than on larger ones. Nevertheless, when comparing the productivity of plots where a certain type of input was used with plots where that input was *not* used, the use of inputs does seem to have a (modest) effect on crop yield. For instance, on plots where chemical fertilizers, chemical pesticides or improved seeds were used, yields per m^2 were higher. It is also possible that the quantities

11 In Tables 4.5, 4.6 and 4.7, eleven cases (plots) had to be deleted. Four of these were outliers, i.e. unrealistically high yields on very tiny plots. In the other seven cases, the respondent had indicated "no harvest", either because s/he did not know or the plot still had to be harvested or the plot had been left idle. Due to the way the data were obtained (hindsight information), the figures should be seen as no more than indications. What matters are the tendencies.

12 The average harvest from the plots in the households' own compounds was twice as high as from the plots elsewhere (0.49 and 0.24 kg/m^2 respectively). This is in line with the average plot sizes: 458 m^2 on compounds and 1750 m^2 elsewhere.

13 For instance, 87% of the plots in the 10-99 m^2 category, 64% of those in the 100-999 m^2 category and 38% of the 1,000+ m^2 category were located in compounds.

of each input used were higher on smaller plots, but this cannot be verified from the data available.

Table 4.6 shows the relationship between the use of material inputs, on the one hand, and land productivity, on the other. As above, inputs have been classified as 'chemical', 'sustainable' and 'irrigation'. The table shows that the use of *more* inputs in general seemed to lead to higher yields. The figures also indicate that the use of one chemical input did not make a difference but that the use of two or three did. The number of sustainable inputs had no direct influence on productivity levels. Finally, and hardly surprisingly, irrigation had a positive effect on crop yields. It should be noted that except for irrigation, the differences were not statistically significant.

Table 4.6 Mean harvest (all crops) by use of material inputs

inputs		N (plots)	mean harvest (kg)	mean plot size (m^2)	harvest per m^2 (kg)
no. of inputs	0-2	52	96	466	0.21
	3-4	68	292	1247	0.23
	5-8	48	536	1103	0.46
no. of chemical inputs[1]	0	85	135	549	0.25
	1	45	248	1153	0.22
	2-3	38	736	1669	0.44
no. of sustainable inputs[2]	0	28	123	426	0.29
	1	57	207	845	0.24
	2-3	83	426	1227	0.35
irrigation	no	92	253	1313	0.19
	yes	76	359	541	0.66

Notes: 1. Chemical fertilizers, chemical pesticides, chemical insecticides.
2. Manure, crop residues, improved seeds/seedlings.
Source: 1999 survey.

Another factor that might account for the high yields of small plots concerns labour. For instance, the smaller a plot, the more frequently the cultivator will be inclined to weed it as it is not so time-consuming. Moreover, it was surprising to find that hired labour was used much more frequently on the *smallest* plots *and* on the *largest* plots than on the plots with in-between sizes.[14] The average harvest from the plots where hired labour had been used was twice as

[14] Namely 42% and 40% respectively, against 19% and 14% on the plots of the 10-99 and 100-999 m^2 categories.

high as that from plots where no additional labour had been hired, even though plots where hired labour was used were on average bigger.[15]

Farming techniques also determine how much is produced. No direct observations have been made but it was recorded who received technical assistance and who did not. Although there were only ten crop cultivators who had received assistance, it was clear that their productivity was almost three times higher than all the other cultivators put together.[16]

Table 4.7 shows mean harvests and land productivity for two household characteristics: household size and the person responsible for crop cultivation. Although, again, differences are not statistically significant, some remarks can be made. One could expect larger households to realize bigger harvests because, in theory, they have more labour at their disposal. However, this is not confirmed by the figures in Table 4.7. The largest households (8+ category) had a relatively low productivity but given the large average plot size in this group and the fact that they realized about the same yields as the 5-7 members category (with less than half the average plot size), their productivity can be viewed from a different perspective. The relatively high yields of the seven single-person households is notable, but is probably more related to the relatively small plot size than to the labour factor.

Table 4.7 Mean harvest (all crops) by household characteristics

household characteristic		N (plots)	mean harvest (kg)	mean plot size (m^2)	harvest per m^2 (kg)
household size	1 member	7	137	179	0.77
	2-4 members	55	257	1157	0.22
	5-7 members	74	256	651	0.39
	8+ members	32	516	1529	0.34
person responsible	male head	42	401	936	0.43
	female head	22	137	688	0.20
	spouse	85	241	1071	0.23
	other h'hold member	17	483	677	0.71

Source: 1999 survey.

15 Average harvests were 0.42 and 0.24 kg/m^2 respectively. Average plot sizes were 1300 m^2 and 836 m^2 respectively.

16 Namely 0.76 and 0.28 kg/m^2 respectively. This difference cannot be explained by the plot size because the average sizes were almost identical: 960 m^2 and 1040 m^2 respectively.

It seems to make a difference as to who is responsible for crop cultivation. In general, men obtained higher yields than women.[17] This is related to the men's greater use of inputs, especially chemical fertilizers and chemical pesticides. Female heads had particularly poor yields (Table 4.7), not only compared with male heads but also with the spouses of male heads, because these spouses tended to cultivate much larger plots. 'Other household members' include children, a brother or sister of the head, or a parent. The fact that they obtained quite high yields might be related to higher labour input.

Despite the finding that average yields were quite low, the perception of most farmers was different. Of the 20 respondents in the crop-cultivating households in the in-depth survey, only two were *not* satisfied with their yields. Most respondents did however admit that yields could be improved. Money is often an important constraint. Several farmers indicated that if they had sufficient money they would be able to irrigate their crops better or buy chemicals, fertilizers and certified seeds or materials to fence their plot to keep out loose animals. The government could come in here by providing credit facilities. One respondent, saying she "could do better", indicated that an alternative way of obtaining credit was to join a farmers' savings and credit group.[18]

Only three farmers said they would be able to improve their yields if they had more space. Others stressed technical improvements on their existing plots, such as crop rotation, the application of manure to increase soil fertility or the use of more chemical inputs. One respondent, David in Lanet, came up with an innovative idea showing his resourcefulness as well as his environmental awareness. By keeping chickens on his plot, he said he could improve his crop yields "by shifting the chicken house to various parts of the plot, so that I can use that part as a seed bed because it has manure from the chicken droppings".

It should be noted that when asked about possible improvements to their urban farming practices, many respondents referred in their answers in first instance to their livestock-keeping activities. Livestock generate more income than crops, which are grown mainly for subsistence. Investing in livestock is seen as more profitable than investing in crop cultivation.

---

17 Namely 0.46 and 0.24 kg/m^2 respectively. Again, the difference cannot be explained by the factor of plot size, being 936 m^2 for the male heads and 992 m^2 for female heads and spouses together (see Table 4.7).

18 See also the section on ECLOF Kenya in Chapter 7.

Problems

Table 4.8 presents the most frequently mentioned problems related to crop cultivation in Nakuru town, as perceived by the cultivators.[19] The respondents were asked to mention any problems they had encountered in 1998 (middle column) as well as their main problem (right-hand column). Although wide-ranging problems were indeed mentioned (see Table A4.4), it is clear that 16% of the cultivators said that they had had no problem. Many problems are not specific to the urban setting in which crop cultivation takes place; for example pests/-insects, destruction by animals, inadequate rainfall, diseases and poor soils. However, the most frequently mentioned constraint is typically urban, namely theft of crops. Mama and Baba Christopher described it as follows (Versleijen 2002: 37):

> Because of the high risk of theft, people only grow crops like *mahindi,*[20] *sukuma* and *maharague*[21] or *kunde*. Then people steal what they need for their own consumption, to take more is useless. If you would start cultivating products like sugar cane, fruits and so on, one would suffer even more from theft, because these can be sold for a good price, so people will start stealing more and more. Therefore, everybody limits his or her crops to those of which you hope that they would not be stolen.

Table 4.8 Most frequently mentioned problems with crop cultivation in town (%; N=160)

	mentioned as a problem	mentioned as the main problem
no problem	16.3	16.3
theft of crops	36.6	24.4
inadequate rainfall	35.0	24.4
destruction by animals	23.8	10.0
pests/insects	22.5	8.8
lack of water for irrigation	12.5	9.4
diseases	9.4	2.5
lack of inputs/capital	6.9	1.9

Source: Annex 4, Table A4.4.

19 See Annex 4, Table A4.4 for a complete list of problems mentioned by crop cultivators.

20 Maize.

21 Beans.

And Mama and Baba Joshua also complained that

> Theft is a big problem. You know, people pass by and they just take what they want. Or those parking boys come and they just take and destroy the crops. And you cannot do anything about it. You know, it is Kenya Railways land so you cannot fence it off and since it is far from home you cannot look after it properly (Versleijen 2002: 53).

As for maize, theft forces people to harvest it before it has fully dried. Drying has to be done at the house, even though there is little space to do so.

Problems like a lack of inputs, a lack of capital and a lack of labour were mentioned by only a few respondents. Three respondents mentioned a very specific problem, namely poor seasonal timing, a sewage burst (that apparently spoilt their crops) and too much rainfall (while, on the other hand, 56 others complained of inadequate rainfall). The answers to the question about which of the problems was seen as the major problem (right-hand column, Table 4.8) are comparable to all the problems mentioned. All types of problems occur in all estates, so there are no clear differences according to housing density. Theft is the one exception. This occurred more often in medium- and high-density areas.[22] For instance, in the low-density area of Milimani theft was not mentioned at all as a problem and provided one has a well-enclosed compound, it is not really a problem. In the words of one respondent living in Ngei Estate:

> This place is very secure. It is enclosed and therefore there is no theft and no harassment from the Municipal Council.

Harassment or destruction of crops by the local authorities is a typical constraint for urban farmers in sub-Saharan Africa, although the practice has decreased over the last few decades as numerous authorities now recognize the importance of the activity for many urban dwellers. According to Nakuru municipal regulations, farming is forbidden within the town's boundaries. The problem for the municipality is that enforcing such rules is difficult and farming in town has thus become a common phenomenon. The municipality allows crop cultivation as long as the crop is less than one metre tall. Maize is thus forbidden, the argument being that thieves and other criminals can hide among the plants. Nevertheless, maize can be seen everywhere and although crop slashing has hardly ever occurred recently, cultivators cannot count on being spared.[23]

The figures presented in Table 4.8 refer to the agricultural season of 1998, which was a fairly normal year in terms of rainfall. As mentioned earlier (Figure 2.1), the situation was quite different in the following year when rainfall – and its monthly distribution in particular – was inadequate. As a result, by far

---

[22] For about a quarter of the crop cultivators in the medium- and high-density areas, theft was the major problem.

[23] See Box 8.1 in Chapter 8 for an example.

the most important problem faced by Nakuru crop cultivators – mentioned by all respondents of the in-depth survey – was drought. That does not mean, however, that other problems were forgotten. Charles, for instance, also complained of "theft, stray animals feeding on crops, and people cutting maize stalks for their animals".

In summary, this chapter has shown that the majority of the plots were located in people's compounds. The average plot measured 1,000 m^2, though sizes varied considerably. Although a wide variety of crops were cultivated, three types dominated: kale, maize and beans. The types of crops grown depended to some extent on the location of the plot and thus with the type of ownership. A wide range of material inputs were used, including chemical inputs and manure as fertilizer. Less than half cultivators irrigated their crops, mostly on plots in their own compounds. Yields were very modest and varied with plot size, the use of irrigation and the person responsible for crop cultivation. Female household heads in particular realized low yields due, among other factors, to their low levels of input use. Finally, the most commonly mentioned problems were theft and inadequate rainfall.

Photo 7 Dairy cows in zero-grazing in Kabachia. Also kale in the foreground and maize and bananas in the background.
(Sam Owuor, 2000)

Photo 8 Chickens kept in a house in Kabachia.
(Sam Owuor, 2000)

5

The animals

Co-author: Samuel O. Owuor

Animal production

Among the Nakuru population, 20% kept livestock in 1998, not only cattle, goats and sheep, but also smaller animals like chickens, ducks, rabbits, doves and turkeys (Table 5.1). By far the most important were chickens, which were kept by over 80% of livestock-keeping households. The percentages of households keeping larger animals like cattle, sheep, goats and pigs were lower, while other smaller livestock besides chicken – like ducks, rabbits, doves and turkeys – were generally even less common. Nevertheless, the numbers of livestock *in the built-up area* of Nakuru town by the end of 1998 can roughly be estimated at 11,600 head of cattle, 6,400 sheep, 6,500 goats, 350,000 chickens, 13,000 ducks, 3,000 rabbits, 1,400 doves and 580 turkeys.[1] These figures are higher than those provided by the local branch of the Ministry of Agriculture (see Chapter 2), except for the number of cattle that were estimated at 25,000 head by the Ministry. The latter can most likely be explained by the fact that the Ministry figures are based on *all* farmers within town, i.e. including those in the peri-urban zone. Nevertheless, it is surprising that all the Ministry's other figures are lower than the survey findings.

---

1 Based on the number of households keeping certain types of animals (column 1 in Table 5.1), the average number of animals (column 2) and the total number of 68,436 households in Nakuru town in 1999 (Kenya 2000). So, for example, the number of cattle is calculated as follows: 26 x 3.9 = 101 head of cattle, x 115 (68,436/594) = about 11,600.

Table 5.1 Livestock in 1998: number of households and 'demography' (averages), by type of animal

type of animal	(1) no. of households abs. (=N)	%	(2) aver. no. at end of 1998	(3) no. born	(4) no. bought	(5) no. died	(6) no. sold
cattle	26	4.4	3.9	1.1	0.5	0.5	0.3
sheep	8	1.3	7.0	0.8	1.8	0.4	0.3
goats	13	2.2	4.4	1.8	0.8	2.2	2.1
pigs	1	0.2	-.-	-.-	3.0	-.-	3.0
chickens	102	17.2	30.1	11.7	38.3	11.8	28.0
ducks	10	1.7	11.4	20.3	0.1	14.8	3.0
rabbits	3	0.5	8.7	6.7	1.0	13.3	0.3
doves	2	0.3	6.0	3.0	1.0	2.0	1.0
turkeys	3	0.5	1.7	-.-	-.-	-.-	-.-

Source: 1999 survey.

Columns 2 to 6 of Table 5.1 provide statistics on the livestock demography for each type of animal. The table shows that although most animals are being reared for both self-consumption and sale, some species are more important as an income source than others (column 6). The other thing that stands out concerns the high death rates (column 5), particularly among the smaller livestock like chickens, ducks and rabbits. Two-thirds of the 121 livestock keepers in Nakuru had experienced at least one death among his/her animals. This applies particularly to chicken: many respondents mentioned typical chicken diseases.

How problematic the keeping of chickens can be is illustrated by the case of Baba and Mama Esther (Versleijen 2002: 63-64). They started of to keeping local chickens[2] that spent the night in the kitchen and roamed around freely during the day. However, as Mama Esther explained,

> (…) they were all stolen. You know, this flat at the back of our house was not there yet so the whole area was bare. What they used to do was to lie down at the fence, put a thread with a needle and a grain of maize and then, when the chickens would eat the maize, they pull it. Since both of us were frequently absent from home, they could easily do that. This is how they stole all my chicken until I remained with one. I ate that one. You know, I thought I can better eat that one than have it stolen as well.

---

2 Local chickens are the chickens that can be seen roaming around in the streets. They are bought at farms and not as young chickens from a company. Compared with 'improved chickens', local chickens have a lower egg production but are more resistant to diseases.

This did not prevent them from starting a chicken business again, albeit more professionally this time:

> We started keeping chickens [again] as a way of giving us some extra security. With chickens you can keep many in a small area if you build cages and keep them above each other. So we bought some 200 young chickens [layers] and started feeding them in order to raise them up to the point they would be old enough to lay eggs. However, it was in that year that Nakuru was hit by a disease among the chickens and all the chicken died before they laid any egg.

Still, Mama Esther remains hopeful that one day she will have more luck with chickens:

> One day I will start keeping chickens again, only this time I would not keep layers, not even broilers, this time I will keep the local chickens. You know, they are cheaper to buy as well as to keep since you do not have to buy much food for them. They can feed on leftovers and find their own food during the day. You just let them roam freely in the compound and in the evening you lock them in a small cage. They are less vulnerable for diseases as well. (...) When I have them, I will train them to stay near the house if we are not there. Hopefully, they will not be stolen then.

In the rest of the analysis a distinction is made between large livestock and small livestock. Large livestock includes cattle, sheep, goats and pigs, and small livestock are chickens, ducks, doves, turkeys and rabbits. It should be borne in mind that 20 of the 121 households with livestock kept both large and small livestock, so there is some overlap between the two categories. Thirteen kept large animals only, while 87 had only small livestock.

Figure 5.1 shows that there is some relationship between the type of animals households keep and the household's income situation. Of the lower-income

Figure 5.1 Types of livestock by household income (N)

	household income (Kenyan shillings/month)	
type of livestock	lower (<10,000)	higher (>10,000)
large	5	8
small	67	20
both	11	9

Source: 1999 survey.

households, only 19% kept one or more of the larger types of livestock. For the higher income group, this figure was 46%. This difference is undoubtedly related to the costs of buying a large animal, cattle in particular. Small livestock are very common among all livestock-keeping households.

The type of livestock kept is not always only determined by a household's welfare level. In some cases, the owner of the land forbids the keeping of animals, an example being Kenya Railways. As Mama Joshua explains (Versleijen 2002: 54),

> We cannot keep any cows, sheep or goats, for a simple reason: Kenya Railways does not allow keeping cattle, sheep or goats in its quarters. So all the animals you see roaming around here are of other people, not of people from Ziwani. If Kenya Railways finds out you are keeping cattle, goats or sheep you might lose them, your house and even your job!

Rearing systems

Data on the rearing systems of large and small livestock are presented in Table 5.2. Animals were either kept on the livestock keeper's own compound or were herded outside ('free range') or a combination of the two ('both'). In one third of the households with large livestock, the animals were only grazing freely in the neighbourhood, while in six others they were partly kept within the compound and partly outside. An example of the latter was Baba Josephine who kept three cows and a calf (Versleijen 2002: 71):

> This year we started herding the animals outside; before it was only zero-grazing. We started herding them just to try, although the system is not as good as the zero-grazing. The yield of the milk is a little bit lower, but zero-grazing is more expensive because you have to buy the grass and it is also a lot of work since you have to feed them and take care of them. Next year, what we will do will depend on the weather. If there is plenty of rain, I will take them out because there is plenty of grass [in Kabachia; DF]. You know, the yield is slightly lower, but the costs are much lower. There may be a bigger chance of diseases, but you know, even with zero-grazing, ticks are coming in with the grass because the grass is just collected outside.

So, it is for purely economic reasons that Baba Josephine decided to practise a mix of zero-grazing and free range. From an economic point of view, this is understandable as feed (grass) is the major cost item in the zero-grazing system.[3]

---

[3] See Table A6.5 in Annex 6.

Table 5.2 Livestock rearing system by type of livestock (%)

(N=; animals)	large livestock (48)	small livestock (120)
on own compound	54.2	45.0
free range	33.3	53.3
both on own compound and free range	12.5	1.7
total	100	100

Source: 1999 survey.

Compared to large livestock, small livestock (mainly chickens) were more frequently left roaming freely (Table 5.2). But keeping chickens in free range is risky, as one respondent explained (Versleijen 2002: 54):

> You know, if you want to keep chickens, that is very risky. First of all, chickens are very likely to die because of the polluted drainage canals from which they will drink. Those canals are polluted by dirt. Second, the risk of theft is very high, since you cannot keep an eye on your chickens each and every moment.

Hence, those who kept chickens as a commercial undertaking either built a special structure for the animals or kept them in their house. It is less risky to keep ducks in free range, at least according to another respondent explaining that they are less vulnerable for diseases (Versleijen 2002: 39). It could be added that the risk of ducks being stolen is also less because they are not generally eaten. But despite the advantages, this is precisely the reason for many people *not* to keep ducks:

> I will never start keeping ducks. You know, ducks are dirty animals and the meat is not good. Actually, I will never eat duck meat. Even the eggs you cannot eat! I know my husband and children would not eat any of them either, so there is no use keeping ducks! (Versleijen 2002: 54)

The reasons why people reared large and small livestock differ. Small livestock were kept first of all for own consumption: almost 60% of those who kept these animals ate most or all of them and another third consumed some of the animals and sold the rest (Table 5.3). Again, this mainly concerned chickens. Those who kept ducks did so for their own consumption, as there is only a small market for these animals. Referring to a household in Ziwani, Versleijen (2002: 38) gives the example of Mama Christopher who kept eight ducks:

> The ducks are used for consumption and sale, although selling is quite rare. Therefore, the ducks can hardly be considered as a source of money. The ducks serve primarily as a food source. The eggs are consumed by Mama Christopher and her grandchildren. Duck eggs are not generally popular, so the main use of the eggs is reproduction. The meat is consumed by the whole household. Overall, duck meat is

not very much appreciated in Nakuru, so it is difficult to sell them in the market. Whereas a hen (chicken) goes for Ksh. 300, a duck sells for 'only' Ksh. 150.

Table 5.3 Purpose of livestock keeping by type of livestock (%)

(N=; animals)	large livestock (48)	small livestock (120)
own consumption only	20.8	49.2
mostly own consumption	6.3	9.2
both own consumption and selling	52.1	31.7
mostly selling	16.7	7.5
selling only	4.2	0.8
hobby/custom	-.-	1.7
total	100	100

Source: 1999 survey.

Large livestock are not consumed very often by the keepers themselves: almost three-quarters of them sold at least some of their animals. However, keeping livestock, either large or small, *solely* for selling was quite rare in Nakuru. One of the exceptions was the household of Baba and Mama Christopher who kept dairy cows and chickens (broilers and layers) purely as a commercial undertaking (Versleijen 2002).[4]

Inputs

Two-thirds of the large livestock keepers in Nakuru town kept their animals partly or wholly in zero-grazing. This concerned cattle almost exclusively and for all these animals grass had to be supplied. Depending on the breed, cows consume 50 to 80 kgs of (green) grass daily.[5] Most livestock keepers do not fetch the grass themselves but purchase it, Baba Josephine is an example (Versleijen 2002: 70):

> During the dry season, the cows are just inside here and I buy grass from the people who sell it. You know, you can often see them passing on their bikes with those piles of grass on the backs of the bikes. They know who has cattle so they go to these people to sell the grass.

4 See also the section on 'Income' in Chapter 6.

5 William Keyah, personal communication, April 2004.

Large quantities of grass are involved, so it offers good business opportunities in Nakuru town. Baba Josephine's grass supplier explained how the system works (Versleijen 2002: 72):

> I cut the grass starting from early morning just along the roadside or at school compounds or areas like that. I have one fixed contract for a guy with cows who I bring eight bags of grass a day.[6] The rest is just loose. You know, I have been supplying Baba Josephine for the past weeks with grass, but we do not have a contract. So if someone else knocks on his gate before me with some bags of grass, Baba Josephine will buy them. You know, he cannot wait for me because what if I have a funeral to attend and will not show up? Then he had no grass to feed the cows. For Kabachia, I know who is having cows and who will buy grass or only feed them outside, so you just go and knock on those gates. Word spreads quick, so if someone buys a cow I will know it the same day when I supply the others with grass, they will tell me.

Table 5.4 shows all other types of inputs the Nakuru livestock keepers used for their animals in 1998. Thirteen (11%) of them had not used any of the inputs listed in the table. These were all small-livestock keepers. In general, large livestock received more attention than small livestock. This applied to all types of inputs, except ethno-veterinary medicines and kitchen remains/food leftovers. Cattle in particular were relatively well taken care of: all cattle holders gave their animals for instance veterinary drugs and feed supplements, while improved breeds/artificial insemination and feeding with crop residues were also very common (77% of the cattle holders).

Table 5.4 Inputs for livestock keeping by type of livestock (%)

	(N=; households)	large livestock (33)	small livestock (108)
no inputs		-.-	10.8
improved breeds/artificial insemination		41.7	5.8
health:	veterinary drugs	87.5	33.3
	ethno-veterinary medicines	8.3	18.3
feeds:	feed supplements	68.8	42.5
	urban waste	25.0	22.5
	crop residues	64.6	37.5
	kitchen remains/food leftovers	-.-	22.5

Source: 1999 survey.

6 One bag of grass weighs between 25 and 30 kgs (William Keyah, personal communication, April 2004).

As seen in Table 5.4, small livestock received fewer inputs than large livestock. This applied particularly to improved breeds and medicines. One of the exceptions was Baba Josephine who kept 200 broilers and 150 layers at the time of the interviews (Versleijen 2002: 75):

> I bought my first chicks from a company called Kenchick. They have an agent here where you can buy chicks. You have to place an order, then they give you the date to come and collect them. You know, it's an old company with a good name, so you know you get good chickens, so I buy them as one-month old chicks. Also my chicken feed I get in town. For the layers I get layer mash and for the broilers broiler finisher or broiler starter. There is only one shop in town where you can buy this. Also the medicine for the chickens I get in town. You know, the diseases are always there but there is a prevention for that. There is always a recommended time you have to give them a dose of a certain type of medicine. We give it to them through the water they drink. We give them no water for some time and then you put the medicine in the water and they start drinking very quick.

In most cases, it is either the head of the household (38%) or the spouse (56%) who is responsible for rearing animals. For large livestock the head and the spouse (both 42%) shared the responsibility equally, but regarding small livestock the spouses formed the majority (57% vs. 37%). The person responsible was in most cases (69%) a woman (which is exactly the same percentage as for crop cultivation). In a few households it was another household member (like a brother or a child of the household head) who looked after the animals, while in five households (4%) a person was employed to do this. In 16% of the livestock-keeping households, taking care of animals was a full-time job for the person involved. In some cases, children were involved as well. For instance, in one household in Lanet, the children did the milking of the cows in both the morning and the evening. In another household (Bangladesh), the children also gave a hand, at least "when not in school".

Moreover, in 22% of livestock-keeping households, additional labour was hired. The large majority of these households kept cows, although there was also an example of someone hiring a herdsboy to find grazing for four goats. In some cases, somebody was hired to fetch grass, usually on a daily basis, as was described above. Payments for these 'grass boys' varied from Ksh. 1,000-1,500/month or Ksh. 20 per bag. One household with four cows in Shabaab employed two labourers, one for milking and one for fetching grass. In another household (Mwariki), the labourer had to keep the pen clean as well, for a monthly payment of Ksh. 1,800. Finally, in a high-income household in Milimani, the spouse was responsible for the animals and she hired somebody to fetch grass, graze the animals, dig and look after the compound, all for a monthly wage of Ksh. 1,500 plus housing, food and toiletries.

Problems

Table 5.5 lists the most frequently mentioned problems encountered by Nakuru livestock keepers.[7] Although 13 of them (11%) said they had not faced any problems, it is clear from the table that animal health is by far the greatest concern for farmers: 72% mentioned it as one of the problems and for 57% it was the major problem. According to one respondent who kept broilers that he sold to hotels in town:

> The business is risky. Once in a while I am forced to clear my stock because of chicken diseases.

Theft came second, mentioned by about 20% of livestock keepers. Given that half of the livestock keepers practised the system of free range, one would expect this percentage to be higher. The above-mentioned example of Mama Esther's chickens was illustrative in this respect. Another respondent, who also kept a few chickens, saw the theft of her chickens not only as relating to the rearing system but also as a form of envy (she also faced the problem of diseases):

> Because I was doing quite well and because of lack of security, one day three of my chickens were stolen. And before I could recover from that loss, six of the chicks died as a result of an outbreak of a disease. Some chicken survived. But because I feared further loss, I started selling and eating the rest.

Table 5.5 Most frequently mentioned problems with livestock keeping by type of livestock (%)

(N=; households)	large livestock (33)	small livestock (108)	all households (121)
no problem	9.1	11.1	10.7
diseases	75.8	71.3	71.9
theft	24.2	20.4	21.5
lack of feed	27.3	12.0	14.9
lack of funds/capital	15.2	10.2	10.7
lack of safe drinking water	24.2	6.5	9.1
predators	3.0	10.2	9.1
lack of space	3.3	7.4	6.6
harassment	12.1	2.8	5.8

Source: Annex 5, Table A5.1.

7 Table A5.1 in Annex 5 details all the problems mentioned and the main problems.

And even ducks, although generally not popular for eating, are not always safe, as Mama Christopher testified (Versleijen 2002: 38):

> The main problem with the ducks is theft. A lot of ducks are stolen by those children on the street who take them home to eat.

Lack of feed and safe drinking water were constraints mentioned by 10% to 15% of the livestock-keeping population (Table A5.1). 'Nuisance', mentioned by two respondents, differs from the other constraints in that it refers more to the farmer's neighbour(s) than to the livestock keeper him/herself. It probably shows that the two farmers mentioning it had problems with their neighbour(s) because of the latter's problem with the farmer's livestock.

Although, generally speaking, the keepers of large livestock and those of small livestock were unanimous regarding the various problems, there are some problems which are more specific to large than to small livestock and vice versa (Table 5.5). Lack of feed and safe drinking water is much more of a problem for large livestock keepers, simply because these animals eat and drink much more than small animals. Harassment, though not frequently mentioned, is also a constraint specific to large livestock. This may be related to regulations which forbid large animals from roaming freely. One problem frequently mentioned by small-livestock keepers was the threat of predators. This is logical, since a chicken or a duck is much easier for some wild animal (or dog) to catch than a goat or a pig.

The same can be said in relation to residential housing density. In general, livestock keepers in the three density categories mentioned the same types of constraints. However, the six livestock keepers in the lowest-density area of Milimani attracted attention by hardly mentioning any problems, except 'diseases' (like all other livestock keepers), 'theft' (one of them) and 'predators' (three). The latter problem is likely to be related to housing density because predators are rarely seen in densely populated areas: only 8% of livestock keepers in the medium-density areas and 5% in the high-density areas mentioned this as a problem.

As with the problems related to crop cultivation in the previous chapter, the problems presented in Table 5.5 refer to the year 1998. The drought of 1999 was not only a major problem for crop cultivators, but also for livestock keepers, particularly those who kept cattle. Fetching grass became increasingly difficult, so either additional food had to be purchased or the animals had to do with less food. This affected the health of the animals (more diseases and more deaths) and milk production.

Waste disposal

Because the large majority of the larger animals are wholly or partly kept within the compound (see Table 5.2), disposing of animal waste is a major concern in town. How did Nakuru livestock keepers dispose of this waste? Table 5.6 shows that one third of farmers said they dumped part or all of the waste in the street. Most of them (92%) dumped the whole lot in fact. However, the table shows that this practice was more common among small-livestock keepers than among those with large animals. On the other hand, many more (62%) of the livestock keepers – either themselves or their neighbours – were able to utilize some or all of the waste productively for crop cultivation purposes. The dung of the larger animals in particular appeared not to be wasted in Nakuru town.

Table 5.6 Disposal of animal waste by type of livestock (%)*

(N=; households)	large livestock (33)	small livestock (108)	all households (121)
use some or all of it for own crop cultivation	78.8	44.4	47.9
give some or all of it to neighbours	18.2	12.0	14.3
dump some or all of it in the street	15.2	34.3	32.8
dump some or all of it in dustbins, pits, compound	-.-	13.9	12.6
sell all of it	-.-	0.9	0.8

* Totals exceed 100% due to combined answers.
Source: 1999 survey.

A comparison between a group of 'pure' livestock keepers (no crop cultivation) and a group of mixed farmers (livestock keeping and crop cultivation) in the separate study by Nyandwaro (forthcoming) confirmed these findings. Most of the mixed farmers used all their animals' waste for their own crop cultivation, while those who had a surplus of waste gave it to crop-cultivating neighbours or sold it. The majority of the 'pure' livestock keepers gave or sold the waste to neighbours, though some also dumped waste in the street.

The re-use of animal waste by livestock keepers themselves or their neighbours was less common in high-density areas (39%) as compared to medium-density areas (69%). Again, the six Milimani livestock keepers stand out as they used 100% of their animals' waste for their own crop cultivation purposes. Dumping waste in the street was equally common in the medium- and high-density areas (35% and 32% respectively). A larger number of livestock keepers in the high-density areas (29%) therefore disposed of their animal waste in dustbins, pits, etc. compared with those in medium-density areas (5%).

In summary, a fifth of Nakuru's population kept either large or small livestock, or both, in town. Large livestock were much more common among relatively well-to-do households, as many of the lower-income households could only afford small livestock (mainly chickens). Animals were either kept in zero-grazing (mainly cattle) or were roaming around in the street (which is illegal). Large livestock were better looked after than small livestock, as shown by the much higher level of input use (especially improved breeds and veterinary drugs). By far the most serious problem encountered by livestock keepers was disease. Finally, as far as the (large amounts of) waste from the animals was concerned, quite a number of the small-livestock keepers dumped it in the street. However, most of the large-livestock keepers used their animals' waste for their own or their neighbours' crop cultivation.

Photo 9 Animal waste disposal site in a compound in Lanet.
(Sam Owuor, 2000)

6

The benefits

Co-author: Wijnand Klaver

In assessing the benefits of urban farming in Nakuru town, different levels and aspects can be distinguished. How much does it contribute to the food supply of Nakuru town as a whole? And how much to the food supply of the households concerned? Does it lead to a better nutritional situation among household members? How much does it contribute to household income? How many people find employment in this sector? These questions are dealt with below.

Importance of urban farming as perceived by respondents

Before embarking on a quantitative analysis of the benefits of farming in town for the people involved, this section starts with a general description of the importance of the activity as perceived by the respondents themselves. First, the crop cultivators and the livestock keepers were asked why they practised this type of activity. For the large majority, the extra food produced was mentioned not only as one of the reasons but also as the main reason (Table 6.1). The household of Mama Esther was a case in point. After she and her husband had suddenly both become unemployed, the vegetables from the compound had become very important for them, as "products are taken from the *shamba* daily for own consumption" (Versleijen 2002: 59). This was, in general, confirmed by Baba Josephine (Ibid: 76):

> Nowadays a lot of people are unemployed. (...) Some small growing of crops or keeping some chickens can be practised by everybody and as such you at least get some food.

Moreover, according to Mama Esther, it not only provides food, it saves money as well with which other food can be purchased (Ibid: 63):

> You can better grow *sukuma wiki* yourself and then buy sugar from the money you would normally spend on *sukuma wiki*.

However, the greens from the *shamba* in her compound could not sustain them for a long period of time, especially not through the dry season. Nevertheless, it was important for them because "as you can grow your own food, then at least you have some security" (Ibid: 64).

Table 6.1 Reasons for growing crops and keeping livestock in town (%)

	crop cultivation (N=160)		livestock keeping (N=121)	
	reasons (total>100%)	main reason	reasons (total>100%)	main reason
food	97.5	90.0	92.6	73.6
income	15.0	3.8	32.2	14.9
to diversify income	9.4	3.8	23.1	10.7
hobby/custom	9.4	2.5	9.9	0.8
other reasons	2.5	-.-	0.8	-.-
total		100		100

Source: 1999 survey.

For a quarter of livestock keepers the additional income obtained from this activity was said to be the main reason. A good example of this was (again) provided by Mama Esther by referring to her neighbour (Versleijen 2002: 64-65):

> My neighbour's husband was unemployed and she got retrenched. With her retrenchment she got a retrenchment pay. From this money she bought a cow and some hens. Now she can keep her children in school from selling the hens and eggs and selling milk.

According to her, this example was not exceptional (Ibid.):

> For most people around, small businesses like this are of extreme importance for the income of the household and the education of the children.

But also in households with a higher income level, the income from livestock in town can be an important means to sustain the household's welfare level:

> Baba Josephine's household would not be able to sustain its livelihood without their urban livestock. (...) They have to secure this income source in such a way that they can sustain or improve their present livelihood. An example of this is that without the practice of urban agriculture, Baba Josephine's son would not make it to the USA. (Versleijen 2002: 77)

Both crop cultivators and livestock keepers were asked about the general importance of their respective activities (Table 6.2). Very few indicated that they "could do without it". For the large majority, urban farming formed at least an additional food and/or income source. And for about a quarter it was a major source. Some even stated that they could not survive without it. Crop cultivation has more of a subsistence nature than livestock keeping. Only 11% of the crop cultivators indicated that this activity was a major or additional income source, against 37% of the livestock keepers. This was well illustrated by the case of Mama and Baba Josephine's household, summarised by Versleijen (2002: 68-69) as follows:

> The cattle provide mainly an income source in cash through the sales of milk and calves, although a small amount of the milk is also consumed by the household. The chickens form mainly an income source in cash as well, although a very small amount is self-consumed. (...) The *shamba* is solely for own consumption and contributes a small amount to the food consumption of the household.

Table 6.2 Importance of urban farming activities for crop cultivators and livestock keepers (%)

	urban crop cultivators (N=160)	urban livestock keepers (N=121)
could not survive without it	6.9	2.5
major food source	18.8	18.2
major income source	3.1	9.9
additional food source	68.1	59.5
additional income source	7.5	27.3
could do without it	3.1	3.3

Note: Totals add up to over 100% because of combined answers.
Source: 1999 survey.

The importance of farming in town for the people involved was well-phrased by Baba Christopher (Versleijen 2002: 39-40):

> People can have several reasons to keep livestock and grow crops in town. First of all, of course it is very important for the food production. Many households here at Ziwani would have severe difficulties if they did not grow their food or kept live-

> stock. (…) You know, for many people it is an important source of surviving. (…) But it is also important because you save money because you do not have to buy food. So that money can be used for other things or you can save.

However, besides these purely economic reasons, Baba Christopher mentioned another reason as well:

> (…) especially here in Ziwani, most people are like me, they come from a rural home. Farming is part of their identity. They have farmed at home and so they farm here in Nakuru, simply because they are farmers. And farming gives them security in a strange surrounding. At least you know for sure you can still provide food for your family. (Ibid: 40)

> You know, we are always farmers, wherever we go we will farm. That is because we grew up with farming and with growing our own food. (…) You know, if you are a farmer once and you are used to grow your own food, you will always look for a way to keep on doing that. That is not only us, that are most of the people in Kenya, simply because most of us are just farmers in their roots. (…) It was not a question whether or not we would cultivate. That was logical. The question was *where* we would cultivate (…). (Ibid: 32-34)

To relate the phenomenon of farming in town solely to economic circumstances would be too simple an explanation. Also in times of economic prosperity, many people indicated that they would grow crops and/or keep livestock. That was confirmed by the answers on the question whether one would stop with farming in town if they had sufficient other sources of income.[1] Only one respondent (out of 24) – who had had trouble with dying animals – said he would stop. All the others would continue, and among them several would even expand their businesses if they had more resources to invest. But despite the 'identity' aspect mentioned above, most respondents gave economic reasons for their decision, such as the one in the low-income area of Rhonda Pondamali who stated that "it helps my family a lot in subsidising our income". Some, however, gave non-economic reasons as well, such as a simple "I like it" or "to cope with idleness". One respondent mentioned the milk from his cow that he would not like to give up.

Whatever the real motive behind farming in town may be, it is without doubt that the *importance* of farming in town has increased considerably due to economic recession, urban poverty and unemployment. Nineteen (out of 24) respondents of the in-depth survey said that compared with when they started farming in town, the activity had increased in importance, mainly because of the food and income aspects. Of the five respondents who said that the importance had *not* increased, four referred to the drought of the year in which the survey

---

[1] In-depth survey 2000.

took place. The other one was the livestock keeper whose animals had died and who lacked space to expand his activities.

Food supply

Food supply at town level

Based on the production figures in Table 4.4 (p. 58), it is possible to assess the contribution of the food produced to the energy requirements of the producers themselves and of the Nakuru population as a whole. It is evident that because of various assumptions on which such estimates are based, these figures can be no more than cautious indications. The calculations – based on the 1998 data – are presented in Table A6.2 (Annex 6). More than six million kg of crops were produced by Nakuru farmers living in the built-up area.[2] If the producers had consumed all of it themselves, it would have constituted almost 30% of their energy requirements. However, an estimated 25% of the produce of the ten main crops in Table 4.4 was *not* consumed in the producers' households but sold instead.[3] In other words, the direct contribution to the producers' energy requirements would amount to about 22%. This also implies that an estimated 1.6 million kg of crops produced within the built-up area of the town were marketed locally. Thus, many other households benefit from urban food production for their food supply, and at prices that are likely to be lower than the normal market prices. In all, the six million kg of crops constituted almost 8% of the total energy requirements of the entire population of the town (Table A6.2). It should be noted that these figures refer to a year that can be classified as 'normal' in terms of rainfall. The following year, 1999, was a comparatively bad year, so the contribution of urban crop cultivation to the town's food supply will have been much lower. Although the necessary data were not available to quantify this for 1999, the qualitative measures in the following section make this quite clear.

Household food security

In order to have a first indication of whether urban farmers were able to reach a higher level of household food security than the non-farmers, two general

---

2 This was calculated as follows. The 160 crop cultivators produced 56,484 kg of crops (see Table A6.2). For the whole sample (594 households), this is 95 kg per household. For the whole Nakuru population (68,436 households), this is then 95 times 68,436 = 6.5 million kg.

3 For the calculation of the percentages 'self-consumed', see Annex 6, Table A6.1. The estimated percentage of self-consumed ranges between 62% for bananas and spinach and 82% for Irish potatoes.

questions regarding this issue were asked to both groups and in both survey years.[4] The results are presented in Table 6.3.[5] As far as 1998 was concerned, the large majority in both groups answered positively ("yes, always") to the question: "Did your household usually have enough to eat during the past year?" Nevertheless, if considered from the 'negative' side, one might also say that among the non-farmers in Nakuru the number of households with food problems in 1998 was twice as high as among farmers (15% versus 7%). Whether this difference can be fully explained by the factor 'urban farming' is doubtful, however. It is more likely to be a matter of differences in household income, as the percentage of very-low-income households was much higher among non-farmers compared to farmers (see Table 3.3 on p. 41).

The picture for 1999 was quite different. In both groups, a third of the households had faced food shortages that year. As for urban farmers, this indicates that their farming in town *does* make a difference: when harvests fail they face food problems. For the non-farmers, the difference between 1998 and 1999 (-16%) must be due to other factors than urban farming. It may well be that food was relatively expensive in 1999 due to the drought (which hit the whole country), so many very-low-income households in this group could not afford to buy all the food they needed. The urban farmers seemed to be more affected (-30%), possibly by the accumulative effect of high food prices (-16%, as for the non-farmers) and harvest failure in urban farming (the remaining -14%).

Table 6.3 Urban farmers and non-farmers: summary of general food security issues (%)

		1998		1999	
		urban farmers (N=209)	non-farmers (N=385)	urban farmers (N=71)	non-farmers N=62)
always food enough?	yes, always	93.3	84.9	63.4	69.4
most important food source last year	urban production and/or other source	45.0	-.-	30.2	-.-
	purchased	36.8	68.1	61.6	90.3
	purchased and rural production	17.7	29.9	8.2	9.7

Source: Annex 6, Table A6.3.

4 There is a methodological problem with this comparison. In both categories a group of rural farmers are 'hidden'. The true 'non-farmers' are those in the non-farmers group in Table 6.3 who do not practice *any* farming, neither urban nor rural. It can be expected that these are the most vulnerable households in terms of food security.

5 For more details regarding the answers, see Annex 6, Table A6.3.

The second question concerned the previous year's most important food source. For many respondents it was difficult to mention *the* most important food source, so combined answers were common. The results in Tables 6.3 and A6.3 (Annex 6) show that even in the 'good year' of 1998, the urban agricultural produce of very few urban farmers in Nakuru town was their main food source. However, almost half (45%) of them felt that urban farming combined with rural production or with purchasing was one of their main food sources. Consequently, fewer farmers depended on purchased food only than non-farmers. Again, the situation in 1999 was worse. Dependency on purchased food was much greater, for both groups.

Tables 6.3 and A6.3 also show that *rural* farming was a source of food as well: 18% of urban farmers and 30% of non-farmers derived about half of their food from their own rural agricultural production in 1998.[6] Moreover, the figures also clearly show that the 1999 drought not only affected agricultural production in Nakuru town but rural production as well. However, the importance of having access to a rural plot is well illustrated by the example of Baba David (Versleijen 2002). His parents had a three-acre plot in Nyeri (185 km from Nakuru), of which one acre was cultivated by his wife who lived there with their four children. The plot was an important food source:

> Baba David does not derive most of his food from the urban *shamba* but rather from his rural *shamba*. The rural *shamba* does not only serve as a very important food source for the moment but also as an important fall-back. If something goes wrong, like an accident which would make it impossible for Baba David to work, or if he gets fired, he knows he has something to sustain his livelihood.

Baba David's rural plot was much more important than his urban plot. First, the urban plot belonged to the railway company, while the rural plot was his own. Second, the rural plot was bigger and produced much more food. Third, because of his job with Kenya Railways, he could be transferred at any moment, thereby losing his urban plot:

> How can I rely on my urban plot when I might be transferred any moment? This is just something small. However, my plot in Nyeri is not something small, we couldn't live without it. And you know, it is always good to have a place to go back to. At least I am sure when something might go wrong, I can feed my family from my *shamba* in Nyeri.

Yet, as Versleijen (p. 46) pointed out,

> the current importance of the urban plot should not be underestimated. Baba David saves a substantial amount of money throughout the year by cultivating a large part

---

[6] For more details on rural farming activities by Nakuru town dwellers, see Foeken & Owuor 2001, Owuor 2002 and Owuor 2006.

> of his own food.[7] If he would not do this, he would have to take one or more of his children from school, travel less frequently to Nyeri or stop other activities [such as] drinking *chang'aa*[8].

Food consumption and nutrition

Food consumption

The main purpose of the October 2000 survey was to assess to what extent urban farmers were better off in terms of food consumption and nutritional status, compared to urban non-farmers. Unfortunately, rainfall in 2000 was even lower than in 1999,[9] so there was very little harvest from urban crop cultivation. This applied not only to those who relied on rain-fed farming but to some extent also to those who normally practised irrigation because during droughts there is an overall water shortage in Nakuru town.

The level and pattern of food intake as judged from the dietary recall[10] was very similar in both groups (see Table A6.4 in Annex 6). Cereal products (mainly maize) constituted 57% of the total dietary energy in both groups and the other staple foods (roots, tubers & starchy staples) added another 5%. The contribution of grain legumes was limited (5% only in both groups). Vegetables were consumed by all households, but fruits were not (only 1 household in 15). Fruit consumption during the two days of recall was relatively more important in the urban farming households (1 in 10 among the urban farmers and 1 in 30 among the non-farmers). There was a small difference in total energy intake: urban farming households consumed about 75 kcal per consumer unit per day more, which can be largely explained as 40 kcal more from grain legumes, 60 kcal more from foods from animal origin and 25 kcal less from oils and fats.

When compared to estimated requirements, energy adequacy stands at 96% among urban farmers as compared to 93% in the non-farming group. These figures point to a modest food deficit. The difference between the two groups is too small to be significant.

At the time of the survey, both groups purchased most of their food (urban farmers 94% and non-farmers 98% of their total energy intake). In those two

---

7 He was living alone in Nakuru.

8 A local (illegal) brew, usually made of maize.

9 Total rainfall was 601 mm (compared with a long-term average of 940 mm) and the 'long rains' failed completely.

10 A dietary recall interview was held covering the 48 hours immediately preceding the start of the interview, thus capturing two days' worth of consumption. For further details, see "Note on consumer units and calculations" following Table A6.4 in Annex 6.

days own rural production contributed some small amounts, which were more diversified for urban farmers (0.4% of energy from beans, vegetables, milk and eggs) than for urban non-farmers (0.2% of energy by way of beans only). However, urban farmers derived on average 100 kcal per consumer unit per day from their own urban production (3.5% of total energy), namely 40 kcal per consumer unit from products of animal origin (milk, chicken, duck and eggs), 23 kcal from cereal products (dry maize and fresh maize), 23 kcal from vegetables (mainly kale, as well as spinach, tomatoes, spider flower and onions), 8 kcal from dry beans, 2 kcal from fruits[11] (papaya and avocado), 1 kcal from Irish potato and 2 kcal from sugar cane.

Although urban farmers purchased most of their food (94% of total energy), it was a bit less than for urban non-farmers (98%).[12] In the pattern of food purchases one can see that urban farming saved buying maize, kale and spinach, milk and potatoes. Urban farmers both produced and purchased more poultry, meat and eggs, legumes and fruits. This resulted in a slightly more diversified diet. On the other hand, urban non-farmers consumed more fish, which may be related to the differences in household income mentioned above (the percentage of very-low-income households being much higher among the non-farmers compared to the farmers). Fish is the cheapest source of animal protein.

Nutritional status

The nutritional status of children aged between 6 months and 5 years was determined by measuring their weight, height and age. As Table 6.4(a) shows, the two groups were affected by under-nutrition, both of a chronic nature (stunted growth) and of recent onset (being thin, referred to as "wasting"). The differences between the two groups show, on average, better height and weight growth among the children of urban farmers. Yet, this is not so clearly reflected in the percentages of children with too low values The results become more meaningful when the youngest age group is singled out (see Table 6.4(b)). Then the advantages of children in urban farming households appear to apply to the

---

11 Note: The quantitative importance of fruits in the diet is somewhat higher than reflected in the figures on dietary energy derived from them because fruits are low in energy content per 100 g compared to dry products such as cereals and legume grains.

12 During the two days of the survey, the following differences appeared: urban farming households consumed less maize (275 g versus 300 g among urban non-farmers), more rice and wheat (80 g versus 60 g), more green grams and dry peas (13 g versus 3 g), more cabbage and spider flower (53 g versus 35 g), less kale and spinach (115 g versus 167 g), fewer potatoes (163 g versus 173 g) and more fruits (10 g pineapple, bananas and papaya versus 1.5 g), more chicken and meat (58 g versus 31 g), more eggs (11g versus 6 g), less fish (6 g versus 12 g) and less milk (160 versus 200 g).

older under-fives (24-59 months). Their median length growth is less retarded in urban farming households than in non-farming households; this results in a somewhat higher median weight-for-age. The same is reflected in the prevalence rates of stunting (moderate), wasting (moderate) and underweight (moderate and severe), which are lower in urban farming households. Among the younger under-fives (6-23 months), the results are the reverse: median height

Table 6.4 Urban farmers and non-farmers: nutritional status of pre-school children*

	children of urban farmers	children of non-farmers
(a) ALL CHILDREN	(N=22)	(N=33)
*In terms of median z-score***		
height-for-age (HAZ: linear growth)	-1.00	-1.29
weight-for-height (WHZ: thickness)	-0.67	-0.59
weight-for-age (WAZ)	-1.13	-1.28
In terms of percentages of children with too low values		
stunted growth (HAZ<-2)	27.3	27.3
wasting (WHZ<-2)	13.6	18.2
underweight (WAZ<-2)	27.3	21.2
In terms of percentages of children with much too low values		
stunted growth (HAZ<-3)	9.1	9.1
wasting (WHZ<-3)	4.5	3.0
underweight (WAZ<-3)	9.1	12.1

(b) BY AGE GROUP	6-23m (N=5)	24-59m (n=17)	6-23m (N=9)	24-59m (N=24)
*In terms of median z-score***				
height-for-age (HAZ: linear growth)	-2.35	-0.90	0.26	-1.64
weight-for-height (WHZ: thickness)	-1.47	-0.63	-1.00	-0.50
weight-for-age (WAZ)	-2.55	-0.97	-0.83	-1.41
In terms of percentages of children with too low values				
stunted growth (HAZ<-2)	60.0	17.6	11.1	33.3
wasting (WHZ<-2)	20.0	11.8	22.2	16.7
underweight (WAZ<-2)	60.0	17.6	22.2	20.8
In terms of percentages of children with much too low values				
stunted growth (HAZ<-3)	0.0	11.8	0.0	12.5
wasting (WHZ<-3)	0.0	5.9	0.0	4.2
underweight (WAZ<-3)	20.0	5.9	11.1	12.5

* Children between 6 months and 5 years of age.

** Legend for z-scores:
- average reference value = 0;
- normal range for individuals is between -2 and +2; below -3 is considered to be severe;
- as the group average drops below 0, the percentage of children below the lower limit of -2 standard deviations increases.

Source: 2000 survey.

growth is much more retarded and median weight for height somewhat lower in urban farming households than in non-farming households. This results in a lower median weight-for-age. The same is reflected in the prevalence rates of stunting (moderate) and underweight (moderate and severe), which are higher among urban farming households, while the rates of wasting (moderate) are similar. The results thus point to the long-term benefits of urban food production, urban farming in past years having had the effect of preventing malnutrition, especially the chronic variety (stunting) among older children. This is all the more remarkable if one considers that the average household income is lower among urban farmers. The lower weight and height values found among younger children in urban farming households, who are nutritionally at the most vulnerable age, may be the consequence of the bad harvest of 2000 (which was an even drier year than 1999) and could be related to the fact that this could not be compensated for by purchasing more food due to the same households' low incomes.

These results need to be interpreted with caution: (i) the sample sizes are very small (and even smaller for the 6-23 months age group): none of the differences were statistically significant; (ii) no information was collected on children's food intake and the link with family intake is not direct due to distribution of food within the household and other factors that influence nutritional status. The survey was carried out in the period of the year when, according to statistics from five clinics in Nakuru town for the previous year 1999, the percentage of malnourished children is at its lowest (in 1999 malnutrition peaked in the period between March and May).

Income

Charles and Rose came to Nakuru in 1965 and have cultivated crops ever since on a half-acre plot in Lanet. They started with kale, but nowadays they cultivate tomatoes, *dhania* (parsley), beans, spinach and kale. They plant twice a year, use chemicals for weeding and irrigate with tap water. The tomatoes and the *dhania* are mainly sold to local middlemen. When the tomatoes are ripe, they harvest about eight crates a week. The *dhania* gave them an income of about Ksh. 18,000 in 1999. In addition, six bags of beans can be harvested each year, of which four are sold. Some of the crops are given away. The remaining beans, as well as the spinach and the kale, are mainly for home consumption. The income from sales is being used to build a house. Charles and Rose are jointly responsible for their farming activities but have to hire labour for planting and picking. The crop residues are used as fodder for the livestock they keep in the same compound (cows, goats, sheep and rabbits, partly for milk, partly for income, and partly as insurance in case they need money) and the animal dung fertilizes the *shamba*. Their farming activities are beneficial throughout the year as they can sell crops and milk all year round. Over the

> years, urban farming has become increasingly important to them, so they would never stop it. On the contrary, they would like to expand.

Charles and Rose offer an example of a successful urban-farming household who are making a reasonable income out of it, directly by selling crops and livestock products, and indirectly by consuming self-produced agricultural products, thus saving money for other expenditures (fungible income). Since most households in Nakuru performed urban agriculture first of all for the food (see Table 6.1), the indirect income was generally more important than the direct income. Baba David, for instance, said he was able to save about Ksh. 1,000 per month by producing part of his food on his urban plot (Versleijen 2002: 44):

> It is better to grow your own food and get something to eat and use your money to send your children to school than to eat your money.

Similarly Mama and Baba Joshua saved about Ksh. 800 per month (Ibid: 52), while Baba and Mama Esther saved from Ksh. 1,000 up to Ksh. 1,500 per month depending on how much they cultivated (Ibid: 59). However, these are savings made in a certain period of the year only. Baba David produced enough to cover six months. In 2000, Mama and Baba Joshua harvested beans sufficient for 3-4 months (1½ *debes*[13]), maize for 2-3 months (2½ bags[14]) and *sukuma wiki* for more than six months. And the same applied to Baba and Mama Esther. Moreover, in years with insufficient rainfall, such as 1999, there may be little or no harvest at all, and hence no fungible income. The loss of (or 'denied access' to) an urban plot may have equally serious consequences, as Baba Christopher explained (Ibid: 35):

> Because I normally grow my own food (...) I can save some money. (...) Now however, because I'm not growing anything in town, I have to use the money. You know, I could get 1½ bag of beans and 3-5 bags of maize from my garden. Say we consume about 2½ kg per meal, that is 5 kg a day and 150 kg a month, so we can live about three months from the garden. Now I have to buy food in those three months. I could have saved a lot of money!

And Owuor[15] gives the following description of Rita's garden:

> In front of Rita's house is a well-tended *sukuma wiki* garden that supplies her household with *sukuma wiki* throughout the year. The *sukuma wiki* is harvested for about four months before planting new ones. During the dry periods, the *sukuma wiki* is watered using tap water. By getting her *sukuma wiki* from this small plot, she saves about Ksh. 25 daily from May to July when there is plenty of rainfall and twice the amount from January to April when it is dry and *sukuma wiki* is expensive. She also

---

13 A *debe* is a measurement comparable with a bucket of about 20 litres.

14 A *bag* contains about 90 kg of dry maize grains.

15 Samuel O. Owuor, field data, 2003.

> plants *kunde* (cowpea) that she uses for consumption in the house, again almost throughout the year. Like *sukuma wiki*, the *kunde* leaves are harvested straight from the *shamba* when needed for consumption.

The direct income aspect was more important for livestock keepers than for crop cultivators. Of the latter, one-quarter said it to be a source of income, against more than half of the livestock keepers (see Table 6.1). For 10% of the livestock keepers, it was even a major income source. Most of these were people selling milk. However, milk production appeared to differ substantially between the various dairy cow owners, ranging from just a few litres to 15 litres per day per animal. Factors determining milk output were, amongst others, the type of animal, the quantity and quality of the feed, as well as the overall health and care of the animals. If well managed, keeping dairy cows could be very rewarding financially.

Baba Josephine may serve as an example of a successful dairy farmer in Nakuru town.[16] An estimate of the profitability of his business is given in Table A6.5 (Annex 6). He had two good dairy cows that provided him with an annual income from milk sales of about Ksh. 180,000. The variable costs – grass, supplementary feeds, veterinary drugs and hired labour – came to about Ksh. 100,000. He therefore made a profit of some Ksh. 80,000 gross. Compared with a total initial investment of Ksh. 75,000 (two cows, a shed and milking utensils), Baba Josephine had good reason to be content with his business:

> Cows are expensive animals to keep, because of the veterinary drugs and check-ups they require. However, they also bring in a lot of money! (Versleijen 2002: 74)

He was even trying to make it more profitable by letting the animals be herded outside by his labourer during periods when there was abundant grass, thus saving substantially on the purchase of grass, which formed the bulk of his variable costs. He even tried to save on veterinary costs:

> The cows I give medicines through injections. I do it myself. Before, we asked the doctor to come, but through experience, through seeing it, I now do it myself. So now I only go and enquire, I tell him the symptoms of the disease and he recommends you a type of medicine. (Ibid: 71)

Milk sales do not provide a constant income throughout the year. Cows produce best during the lactation period, i.e. for about 210 days per year. How much they produce during the rest of the year depends largely on how well fed they are. Particularly in a dry year like 1999, milk production could go down substantially; an extreme example being a household whose yield dropped from 22 litres per day during the lactation period to just one litre during the dry season.

---

[16] See Versleijen 2002: 70-74.

Chickens can be another rewarding business. Broilers are easy to raise from young chickens. It only takes seven to eight weeks for them to be fully grown and then they can be sold at a good profit. Layers can be equally rewarding. One respondent living in an estate called Shabaab kept chickens that produced on average 60 eggs per day. She sold these for Ksh. 5 each. On an annual basis, that comes to over Ksh. 100,000. Her costs were about Ksh. 50,000 for commercial feed, vaccinations etc., so she was able to make a reasonable profit. Her main difficulties were diseases and drought (in 1999). Two other respondents mentioned the same problems. One of them had to eliminate the whole flock to avoid the disease spreading. The other was able to sell the dead animals, albeit for a price about half of that for a healthy chicken.

Marketing was usually a very simple affair. In most cases the milk and/or eggs were sold to people in the neighbourhood. Some producers had built a small kiosk on the fringe of the compound facing the street, where they sold their products. However, there were also farmers who sold to retailers and/or wholesalers. Baba Josephine, whose two cows produced almost 30 litres of milk per day during the lactation period, sold to both types of traders (Versleijen 2002: 73):

> I'm selling to the *duka*[17] who then sells it to people, or I sell it to Asians who then sell it to retailers. I might make more money to sell directly to consumers myself but then I have to stay at home the whole day and store the milk during the day. Now I sell everything in the morning and I am through. And I am also sure I can sell all my milk before noon so it doesn't get spoiled.

In some cases, chickens and eggs were sold to traders too. Again, Baba Josephine offers a good example. At the time of the study, he had 200 broilers. Selling them from the house was not feasible because

> The supply of chickens is high, there are many people selling them. So, what you have to do is go to the restaurants and look for orders. (Versleijen 2002: 74)

After a hotel or restaurant had placed an order, the chickens were slaughtered at Baba Josephine's house. The eggs from his 150 layers were sold to a wholesaler in town (see below).

Employment

It is not easy to assess the importance of urban agriculture in Nakuru town in terms of employment. However, with the data available from the 1999 survey, a cautious estimate can be made of the number of people working in the sector.

---

[17] Kiosk.

These figures are presented in Table 6.5. Based on a total number of 68,436 households in Nakuru at the time of the survey (Kenya 2000), some 24,000 were involved in urban agriculture. Of these, about 18,400 could be labelled as crop cultivators and 13,700 as livestock keepers (see Table 3.2). More important, however, is the figure of about 4,900 people who reported their farming in town to be a full-time job. In addition, in about a quarter of both the crop-cultivating and the livestock-keeping households, hired labour was used. In other words, for about 8,400 persons urban farming constituted a form of paid labour. For these people, working in urban crop cultivation was an income-generating activity, albeit of an irregular and seasonal nature.

In all, an estimated 13,000 persons in Nakuru found employment in the sector in 1998, either full-time or part-time. Compared with the total Nakuru labour force of 141,181 in the same year (MCN 1999: 62), one can only conclude that in terms of employment, urban agriculture constitutes an important sector.

Table 6.5 Employment in the urban agricultural sector[a]

	crop cultivation		livestock keeping		total
	%	N	%	N	N
households involved (% of total sample)	27[b]	18,900	20[b]	14,000	24,650[c]
- full-time job	14[d]	2,650	16[e]	2,250	4,900
- hired additional labour	28[d]	5,300	22[e]	3,100	8,400

Notes: a) The percentages have been extrapolated to Nakuru town as a whole based on a total number of 68,436 households in 1999 (Kenya 2000). Because of the speculative nature of these estimates, figures have been rounded off.
b) Percentage of total sample.
c) Since a number of households are involved in both crop cultivation and livestock keeping, the total is lower than the sum of the previous columns (see also Table 3.1).
d) Percentage of urban crop-cultivating households.
e) Percentage of urban livestock-keeping households.
Source: 1999 survey.

Other people benefit indirectly from the sector as well, namely suppliers of inputs and buyers of the produce. As for the supply sector, the earlier-mentioned 'grass boys' were a case in point. It is impossible, however, to assess how many were active at the time of the study. Other suppliers were those who sold supplementary feeds and medicines related to livestock keeping, and seeds, chemical inputs and tools in the case of crop cultivation. No data are available on the numbers of such suppliers, but it can be assumed that there were only a few. As mentioned in Chapter 5, according to Baba Josephine, there was only one shop in town where he could buy his chicken feed.

As for the buyers of farm produce, this usually involved milk and eggs. An example of a milk trader was a man called Mkamba who bought (some of) Baba Josephine's milk:

> Each morning I go to Kabachia by bike to collect five litres of milk. I pay Baba Josephine immediately. The reason I buy from him is that he sells me milk that is of high quality, which means that he never adds any water, Blueband or wheat flour to the milk. You know, if you have good milk, you can make much more tea out of it than when you have a lower quality of milk. Right now, customers start coming to my *duka* because they know my milk is of good quality. You know, like that I can attract many customers. (Versleijen 2002: 72)

And the possible success of the egg business is shown by a lady called Murugi, who bought Baba Josephine's eggs:

> I started buying and selling eggs as a small business when I got retrenched. By that time, I went to look for people who had eggs, bought them and sold them. However, the business has grown and these days I have a name and people come to me to sell the eggs, so now I can regulate everything from my little office here. Eggs are brought by the farmers on a weekly basis. It does not matter how many a farmer brings, I will buy any number. This is a relationship you build up with your suppliers; they can rely on me that I will always buy. Only if the supply is higher than the demand I have to turn people down; then I only buy from the suppliers that are already there for a long time, like Baba Josephine. Eggs from Nakuru are cheap compared to other areas and the quality is good. They are sold all over Kenya and also to Tanzania and Uganda. (Ibid: 75)

In summary, the benefits of urban farming for the people involved are manifold. It provides households with food that would otherwise have to be bought (indirect income). Its importance as a food source was for instance shown by the better nutritional condition of children (compared to children of the non-farmers) and the food problems faced by these households when harvests fail due to drought. For many households, urban farming is an income source as well. Livestock keeping can be quite rewarding, as some examples in this chapter have shown. Finally, farming appeared to be an important sector in terms of urban employment.

7

The support

Co-authors: Samuel O. Owuor & Peter W. King'ori

This chapter consists of three parts. The first part deals with some general notions on assistance for crop cultivators and livestock keepers in Nakuru town. These findings are based on the general survey of 1999. The second part presents the main results of a study in which a group of urban farmers supported by the Agriculture and Rural Development Programme (ARDP) of the Catholic Diocese of Nakuru is compared with a group of unsupported urban farmers in terms of agricultural and economic issues. The third section deals with the recently launched Ecumenical Church Loan Fund (ECLOF Kenya) and presents examples of successful micro-credit assistance for low-income urban farmers.

Assistance for crop cultivators and livestock keepers

Crop cultivators

Only ten (6.3%) of the 160 crop cultivators received technical assistance in 1998. Three had been visited by an extension officer and one by an officer from the Agricultural and Rural Development Programme. The others had received assistance from neighbours and/or relatives. It is notable that these ten crop cultivators realised a much higher productivity on average than those receiving no assistance at all (see Chapter 6).

Receiving assistance occurred somewhat more frequently on larger plots (perhaps they were more visible to the extension officers) and on plots only recently given over to cultivation. The latter might be related to the finding that younger crop cultivators as well as more educated ones received more assistance than older cultivators and less-educated farmers. Perhaps the young and relatively well-educated are better able to find ways of gaining assistance than the older and the less educated.

Livestock keepers

Technical assistance among livestock keepers was more widespread than among crop cultivators: 30% of the former had received some assistance in 1998. Assistance for keepers of large livestock was much more common (55%) than for those with small livestock (25%). Assistance was mostly provided by a professional officer (39%), a neighbour (25%) or a combination of an officer, neighbour and/or relative (17%). Of those households who did receive assistance, over half (53%) had been visited by an extension officer. On the other hand, if taking *all* livestock keepers into account, only 19 (16%) had received any assistance from the Ministry of Agriculture. One farmer had received assistance from the Catholic Diocese of Nakuru's urban agriculture programme.

There appeared to be no relationship between receiving technical assistance, on the one hand, and the number of animal deaths, on the other. Among those urban livestock keepers who received assistance and those who did not, the percentage of farmers who experienced one or more deaths was equally high (67% and 64%, respectively). This may be due to the fact that the vaccination of animals was usually only done *after* the outbreak of a disease. Preventive measures were not practised.

The data presented in Table 7.1 show that it not only made a difference as to whether an urban livestock keeper received assistance or not, but also from whom he/she got support. A distinction was made between professional officers (extension workers from the Ministry of Agriculture, programme officers and veterinary officers) on the one hand, and neighbours and/or relatives on the other.[1] As far as rearing system was concerned, there were no big differences between the three groups (Table 7.1). Nevertheless, zero-grazing ('within compound') was more common among those who received professional support than in the other two groups. Consequently, leaving one's animals to roam freely around outside one's compound was less common in this group.

Nearly all livestock keepers used some kinds of input. However, those who did not use any input were to be found in the group without support. In general, the percentage of livestock keepers using certain specific inputs was highest among the professionally-supported farmers, somewhat lower among those who received assistance from relatives or neighbours and much lower in the non-supported group (see Table A7.1). This shows first of all that supported live-

[1] More detailed figures are presented in Annex 7, Tables A7.1 and A7.2.

stock keepers used more different types of inputs than the other groups. For some types of inputs, the differences were remarkable (Table 7.1). The use of improved breeds and veterinary drugs was common among the professionally supported livestock keepers, but clearly less so in the two other groups. Feed supplements were very commonly given to the livestock of the two supported groups, but less often in the non-supported group. To a lesser extent, this applied to the use of crop residues as feed as well.

Table 7.1 Summary of livestock-keeping characteristics by type of assistance (%*)

		received assistance?		
	(N=)	yes, from professional officer** (25)	yes, from neighbour or relative (11)	no (85)
rearing system	within compound	52.0	36.4	38.8
	free range	32.0	45.5	55.3
types of inputs: % yes	improved breeds	68.0	18.2	8.2
	veterinary drugs	96.0	45.5	36.5
	feed supplements	92.0	81.8	47.1
	crop residues as feed	68.0	63.6	48.2
disposal of animal waste	use part or all for own crop cultivation	68.0	36.4	43.4
	dump part or all in the street	20.0	27.3	37.3
types of problems: % yes	diseases	76.0	81.8	69.4
	theft	12.0	9.1	25.9
	lack of feed	36.0	9.1	9.4
	lack of funds/capital	12.0	9.1	10.6
	lack of safe drinking water	20.0	18.2	4.7
	harassment	12.0	9.1	9.4

* Percentages are from 25, 11 and 85 cases, respectively.
** Includes extension officers, programme officers and veterinary officers.
Source: Annex 7, Tables A7.1 and A7.2.

Receiving assistance from professional advisers also seemed to have a positive effect on the ways livestock keepers disposed of the waste from their animals. In general, this is one of the major environmental problems related to keeping animals in town. As shown in Table 7.1, the professionally supported livestock keepers were more inclined to use their animals' manure for their own crop production. At the same time, they dumped the waste in the street less frequently. However, one in five of them still practised this unhealthy habit.

Table 7.1 also shows the most frequently mentioned problems related to livestock keepers' activities.[2] Although the three groups did not differ substantially in this respect, some differences can be noted. For instance, supported farmers had relatively fewer problems with theft of their animals. This is likely to be related to the rearing system, since zero-grazing was more common among them – and particularly among the professionally supported farmers – than among the non-supported farmers. The professionally supported livestock keepers complained relatively more often of a lack of feed. It is not totally clear why this was so, but it might again have to do with the rearing system because zero-grazing implies that feed has to be collected daily from elsewhere. Finally, a lack of safe drinking water seemed to present no problem at all for the non-supported livestock keepers, but for a fifth of the supported ones it did. Again, why this should be so is not entirely clear. An explanation might be that supported farmers were more aware of the quality of the water their animals drank, which in turn might be a result of the support officers' advice.

The Agriculture and Rural Development Programme (ARDP)

The Catholic Diocese of Nakuru (CDN) has been involved in agricultural development since 1974. The current Agriculture and Rural Development Programme (ARDP),[3] which emerged from the Drought Rehabilitation Programme set up in 1985, came into existence in 1992 and had the following objectives:[4]

- to encourage rural communities to be in charge of their own development (self-reliance);
- to provide the necessary support to such communities to attain their own development; and
- to increase the sustainability of community/group initiatives or projects.

To achieve these objectives, the programme emphasizes "the mobilization and organization of homogeneous farmers' groups/communities, which will be able to identify and analyse their own problems and needs and consequently seek solutions which can be reached primarily through self-help and in case of outside help, this should only be to facilitate this self-help local initiative in realizing their mission" (ARDP 1998: 2). More concretely, the programme's activities include: training and education; soil and water conservation; crop production and management (food security); and livestock production and

---

2 See Table A7.2 for all mentioned problems.

3 In 2004, the programme faced problems causing the programme coordinator to resign. It is not clear whether activities have meanwhile been resumed.

4 The general information on ARDP is obtained from the programme's Quarterly Report 2, 1998.

development. The programme covered five districts,[5] with the target population being small-scale farmers including some urban farmers in Nakuru town.

A special Masters-level study was carried out to assess how far urban farmers in Nakuru benefited from the programme in terms of socio-economic status, food consumption and the nutritional condition of the children and their mothers (see King'ori forthcoming).[6] A group of ARDP farmers was compared with a group of non-participating (or non-ARDP) farmers.[7] The group of ARDP farmers consisted of 29 households, 14 of whom were 'true' urban farmers, i.e. living and farming within the municipal boundary of Nakuru town, while the other 15 were located just outside the municipal boundary (but within the Nakuru Planning Zone). The group of non-ARDP farmers consisted of 48 households, selected for being close neighbours of the ARDP farmers.

The 29 ARDP farmers received various kinds of support (Table 7.2). Training – or indirect support – was the most common type of support provided to the Nakuru farmers. This included sustainable agriculture, agro-forestry, crops and livestock management, community organisation, leadership skills and planning and monitoring skills. Regarding direct support, six (20%) of the ARDP farmers received help in the construction of a water tank, allowing them to practise some form of irrigation in their small vegetable gardens during the dry season. Five farmers received assistance with the construction of a zero-grazing

Table 7.2 Forms of support for the ARDP farmers (N=29)

Training	21
Water tank	6
Zero-grazing unit	5
Loan for livestock keeping*	5
Beehive	4
Loan for crop cultivation**	3
Provision of dairy cow	2

* To buy chicken/rabbits, drugs, feed.
** For cereal marketing, fertilizer, seeds.
Source: ARDP study, 2000.

---

5 Nakuru, Kericho, Bomet, Koibatek and Baringo.

6 Here, however, mainly agricultural and economic issues are dealt with.

7 Obviously, the more appropriate way to assess to what extent ARDP farmers benefited from the project would be to do a longitudinal study to compare the ARDP farmers' situation before the start of their participation in the project and how they did afterwards. However, this was not possible within the context of a Masters study.

unit and another five received a loan to buy livestock and/or livestock-related inputs. Four farmers were supplied with a beehive on credit, while two received a dairy cow as an investment in commercial milk production as well as for self-consumption. Finally, three farmers received a loan related to crop cultivation, for instance to buy inputs or for maize marketing. In general, support for livestock-keeping activities was more common than for crop cultivation.

Both groups were predominantly mixed farmers, combining crop cultivation and livestock keeping. The average plot size for crop cultivation among the two groups was about the same (6,800 m^2 and 7,400 m^2, respectively). Cultivation methods were also the same among the two groups, which is, for instance, shown by the use of material inputs.[8] The use of improved seeds/seedlings and chemicals for crop cultivation was very common in both groups. The same applied to the use of manure as fertiliser, which is obviously related to the fact that most of these farmers kept livestock as well. Irrigation was not common. The only difference between the two groups concerned extension services for crop cultivation, which were more common among ARDP farmers. As for inputs for livestock, ARDP farmers used 'modern' inputs/techniques such as artificial insemination and veterinary drugs somewhat more frequently, which can be related to their participation in ARDP.

There were differences between the two groups as far as problems with crop cultivation and livestock keeping were concerned.[9] The supported farmers complained less about 'lack of water' and 'crop pests and diseases', but somewhat more about the 'poor market' for their crops. As for livestock keeping, ARDP farmers had more frequent problems with obtaining enough fodder for their animals. They also complained somewhat more frequently of 'lack of space' and 'high prices of inputs'. The latter can be related to the higher use of 'modern' inputs. Yet, they did not complain of a 'lack of capital', as quite a number of the non-ARDP farmers did.

To get a crude picture of the welfare level of the two groups, two indices were constructed. The *welfare index* consists of the ownership of such items as a television, radio, sofa and bicycle (and in a few cases even a vehicle). The *house quality index* is a measure of the quality of the materials used for the floor, walls and roof. The aggregate average scores are presented in Table 7.3. The figures show that there were few differences between the scores of the ARDP farmers and those of the non-ARDP farmers.

---

8 Figures on the use of inputs for both crop cultivation and livestock keeping are presented in Annex 7, Table A7.3. Note that the use of inputs for crop cultivation was much more common among these peri-urban farmers than among the farmers in the built-up area of Nakuru town (see Chapter 4, Table 4.2).

9 The figures regarding the most frequently mentioned problems are given in Annex 7, Table A7.4.

Table 7.3 ARDP farmers and non-ARDP farmers: material welfare levels

	ARDP farmers (N=29)	non-ARDP farmers (N=48)
welfare index*	40.5	43.9
house quality index**	5.5	5.2

* Vehicle = 100; television = 15; radio = 6; sofa set = 5; bicycle = 4. Maximum score is 130.
** Construction materials of floor, walls and roof. Maximum score is 9.
Source: ARDP study, 2000.

As mentioned above, the focus of the programme was more on livestock than on crop cultivation and more particularly on cattle and milk production. For that reason, Table 7.4 presents figures concerning the sale of cattle and milk.[10] First of all, the percentages of households selling cattle and milk were higher in the project group. The average number of cattle and the average amount of milk sold were also higher in the ARDP group. As a result, the average gross income from sales of cattle and milk was much higher among ARDP-supported farmers.[11]

Gross income from crop sales was about the same in both groups.[12] Hence, in sum and in agricultural terms, the ARDP farmers as a group seemed to fare better than the non-ARDP farmers (i.e. when outliers are excluded). In terms of farm management, the project group seemed to perform better as well, indicated for instance by the higher use of modern inputs in this group. This can likely be related to training and education, which was by far the most common type of assistance provided to the participating farmers.

---

10 Sales of other types of livestock (sheep, goats, pigs, chickens and rabbits) are omitted here because in all cases (except sheep) only a few households were concerned, while just one household accounted for nearly all the animals sold. For instance, two households sold a total of 14 pigs, of which 13 were from one household. In the non-ARDP group, four households sold a total of 405 chickens, of which 400 were from one household. As for sheep, these were sold by seven and four households in the ARDP and non-ARDP group, respectively, with an estimated average gross income (at group level) of Ksh. 1,618 and Ksh. 340, respectively. As for eggs, none of the ARDP farmers sold any, but six of the non-ARDP farmers did. Two of the latter were egg 'giants' who sold an estimated 12,700 and 53,100 eggs in the year preceding the survey.

11 At least when the two 'outliers' regarding milk sales in the non-ARDP group are excluded (see footnote in Table 7.4). If included, the average gross income from the sales of cattle and milk would be about the same in the two groups.

12 Namely Ksh. 4,530 and Ksh. 4,630, respectively.

Table 7.4 ARDP farmers and non-ARDP farmers: cattle and milk sales

	ARDP farmers	non-ARDP farmers
Cattle	*(N=29)*	*(N=48)*
- % households selling cattle	24%	13%
- average number of cattle sold	0.4	0.2
- average gross* income from cattle sales (Ksh.)	4,508	1,816
Milk	*(N=29)*	*(N=46**)*
- % households selling milk	41%	31%
- average amount of milk sold (litres)	371	261
- average gross* income from milk sales (Ksh.)	8,422	5,918

* Without deducting production costs.

** Two households selling disproportionately large amounts of milk (5,250 and 7,500 kgs) have been left out here. If included, average gross income from milk sales for the non-ARDP group would be Ksh. 11,752.

Source: ARDP study, 2000.

The Ecumenical Church Loan Fund (ECLOF Kenya)

The Ecumenical Church Loan Fund (ECLOF) is a global initiative that has its headquarters in Geneva, Switzerland. The Kenyan branch – ECLOF Kenya – was launched in 1994. Besides the country's head office in Nairobi, ECLOF Kenya has offices in six other towns, Nakuru being one of them. ECLOF Kenya supports the building of sustainable communities by providing fair credit services for human development in both rural and urban areas. One of its main objectives is "to increase accessibility to credit by the economically active and marginalized micro/small business and farming people of Kenya".

Set up in 2001, the Nakuru town office served about 600 members (clients) three years later, most of whom were small-scale traders.[13] Only a small proportion (5%) of the members were farmers, benefiting from ECLOF's financial assistance for expanding or improving their farming activities. The large majority of these farmers engaged in dairy farming (zero-grazing) and poultry keeping in the rural areas. Five of the farmers could be classified as *urban* farmers, i.e. living and farming within the municipal boundary or, more precisely, in the peri-urban areas of Nakuru town. Due to the requirement that the activity must be income generating (see below), all five urban farmers engaged in livestock keeping: three in dairy farming (zero-grazing), one in pig keeping and one in poultry keeping. Four of these cases were interviewed and are described below.

13 Second-hand clothes, retail shops, *matatu* businesses, chemists, tailoring, greengrocers, hawking, etc.

Membership to ECLOF Kenya is open for anyone, provided that the individual members go through the pre-designed registration procedures and requirements. There are no special conditions for members who want to improve or expand their farming activities. First and foremost, potential members must be engaged in an income-generating activity (business or farming) and at the same time belong to a registered group. This is usually a group of friends or neighbours with a common interest. Members of a group who are well known to each other are, therefore, able to co-guarantee one another when applying for a loan.[14] While ECLOF Kenya encourages already-existing groups, a large majority of their members came together after learning about ECLOF's activities.[15]

Once prospective members have formed a group and registered it with the local Department of Social Services, each will pay a non-refundable registration fee. The registration fee is part of the 3% loan-processing fee members are charged. Thereafter, all the members of the group must attend (together) weekly training sessions for a period of eight weeks. The course aims at counselling the members and passing on the necessary business and savings skills. It also serves as a platform for the members to know each other well, to develop a sense of trust and to make the group more cohesive. During this period, each member is required to save, in the group account, 20% of the money s/he has applied for.[16] Ascertaining and valuing applicants' securities is also done during this period. Securities are in the form of a list of items the applicant gives and their equivalent values, i.e. land title deeds, a radio, television set, etc.

ECLOF Kenya's *Ordinary Jiwezeshe Credit Scheme*[17] for registered groups with at least 10 but no more than 30 members is popular with farmers in Nakuru.[18] In this category, the members can access loans of between Ksh. 5,000 and Ksh. 150,000 each, depending on the loan cycle. However, it is very rare for members to apply for over Ksh. 50,000 because of the one-year repayment period, at least for the first two loans. The aim of the loan is to expand or im-

---

14 An example from yet another loan scheme, Pride Kenya, was the Baraka women's group, consisting of 36 members. Each member contributed Ksh. 500 per month, with which members facing a crisis situation – such as death, sickness or the inability to repay the Pride Kenya loan – could be assisted (Samuel O. Owuor, field notes 2003).

15 ECLOF Kenya occasionally carries out outreach and promotional activities in its regions of operation.

16 ECLOF Kenya is a mandatory signatory to all the group accounts.

17 *Jiwezeshe* is a Kiswahili word literally translated as "enable yourself".

18 There are other loan schemes that fall outside the scope of this discussion. For example, loans for school fees (for all the members), small group loans (for four to five individuals with a credit or loan history and common-bond activities) and institutional loans (church, schools, etc).

prove on an existing income-generating activity, i.e. meeting working capital, farm inputs and moveable assets needs.

After the loan is granted, ECLOF Kenya maintains monthly follow-up meetings with their members to evaluate or re-evaluate the income-generating activity and repayment progress. If a member is unable to repay his/her loan, the group holds a meeting to discuss the problem. Normally, there are two options: (1) to talk to the member about continued repayments and/or (2) to help him/her pay. It is only as a last resort that the securities are sold.

Case 1: Baba and Mama Fred – diversifying sources of income

Baba and Mama Fred[19] lived in a one-room mud house in Mwariki with their three children. Mwariki is a large, low-income neighbourhood in the south-western part of Nakuru municipality. Their house was built on a half acre plot Baba Fred purchased in 1995. Mama Fred cultivated *sukuma wiki* (kale), spinach, tomatoes and onions on the intensively utilized plot. Besides the crops, both Baba and Mama Fred kept chickens. At the time of the interview, they had about 10 *kienyeji* (local) chickens.

Baba Fred came to know about ECLOF Kenya through a friend and later discussed it with his wife. The only information he was given by his friend was that "there is an organization in town giving financial assistance to small-scale businessmen and farmers". After discussing it with his wife they decided to "try their luck" and after he got the courage required, he went to ECLOF's office in Nakuru town. He was encouraged by what he was told and immediately started to mobilize other people in the neighbourhood, so as to form a group and be able to apply for loans from ECLOF. In less than a month, Baba Fred had formed a self-help group with 20 members.[20] This was in the first quarter of 2002. Given that only one household member can apply for a loan, there was no need for Mama Fred to join the group. After undergoing the ECLOF's training programme Baba Fred received the Ksh. 40,000 he had applied for.

Baba Fred said that he applied for the loan to buy a "grade"[21] cow so that he could start selling milk while at the same time continuing with poultry keeping. True to his dream, he bought a Friesian from another farmer in Kiamunyi, just beyond the municipal boundary. Out of the loan, he paid Ksh. 35,000 for the cow while the rest was used to transport the cow from Kiamunyi to his plot in Mwariki. The cow, which gave birth in 2003, produced on average about 12

---

[19] As explained earlier, Baba and Mama Fred means father and mother of Fred, the first-born.

[20] Not all of these members were farmers like Baba Fred. That applies to the other three cases as well.

[21] This is a crude translation for a "high breed" cow.

litres of milk per day for a period of up to eight months in a year albeit with fluctuations depending on how the animal was fed. Out of the daily production of 12 litres, they sold 10 litres of milk at Ksh. 20/litre and consumed the rest themselves. After buying the cow, Baba and Mama Fred earned an extra, gross income of about Ksh. 6,000 per month from sales of milk. Baba Fred estimated his total production costs at about Ksh. 2,000 per month, which included spraying, vaccinations, buying Maclick, water and sometimes Napier grass. Hence, Baba Fred was glad to "end up with something every month". To reduce production costs, they planted a small stretch of Napier grass outside the gate.

Mama Fred admitted that with the income from the sale of milk they were comfortably able to take care of their school-aged children, buy other food items and even help their parents at the rural home once in a while. Mama Fred concentrated on farming the plot while Baba Fred found a job as a shop attendant in town after a period of unemployment. Before coming to Nakuru with his family in 1991, Baba Fred was a businessman in Nairobi. He had a shop in Nakuru until 1996 when he stopped to concentrate on poultry keeping. Baba Fred confirmed that life was not all that easy before he bought the cow because the profits from the shop and later from his poultry were not that good. His major problem was poultry disease that occasionally hit his flock. However, he consoled himself with the fact that they were lucky because they had access to a plot that gave them food most of the time, i.e. daily *sukuma wiki* or spinach.

Baba Fred considered the profit from the milk as the household's main source of income because (1) with good animal husbandry he was always assured of a profit and (2) he was not sure about staying in his present casual job. In addition, they continued to receive additional income from the sale of chickens and eggs. Baba Fred said that he was gradually improving his skills in poultry keeping.

They both had a very high regard for ECLOF Kenya and the staff in Nakuru who "visit them every now and then". Baba Fred had finished repaying his loan and was considering taking out another. Though he never had a problem with his repayments, he admitted that during the first twelve months most of the profit went into his repayments. He was next planning to plaster the floor and walls of his house. Even with all this happiness, Baba and Mama Fred were not happy with the Municipal Council of Nakuru which constantly harassed them for keeping animals in town and "yet they do not understand how it helps the people involved".

Case 2: James Mwangi – improving and expanding pig farming

James Mwangi first heard about ECLOF Kenya in 2002, a few months after he started keeping pigs in Langa Langa estate in Nakuru municipality where he had access to about an acre of land. Besides the pigs, James had two cows that provided him with milk to sell and for home consumption. He recalled that he was particularly impressed by ECLOF's loan system with its monthly repayments, unlike other micro-finance institutions in Nakuru where this was demanded on a weekly basis. Following advice from ECLOF Kenya, James and his friends registered a group of 19 members and applied for loans. James applied for Ksh. 30,000 to improve and expand his pig-farming activities. The loan was granted in 2002 after he had attended the mandatory training sessions.

The first step for James was to "increase his herd of pigs". In addition, he also renovated one of his buildings and turned it into a butchery where he could sell pork. The pork was from his stock but sometimes he bought from other farmers the same way he sold his to others. By the time of the interview, he had about 20 pigs in his sty. James said that the number was higher than what it was because of their high reproduction rate. Due to a lack of food, he was forced to sell some of them. He explained that reducing the number of pigs was good practice "because it can be very expensive to feed a large stock." According to James, the price of a pig ranged from Ksh. 3,000 to Ksh. 5,000.

Photo 10 James Mwangi's butchery.
(Sam Owuor, 2004)

The business-minded James was glad to be making between Ksh. 10,000 and Ksh. 15,000 per month in profit from his pig business. After repaying his loan without a problem, James was planning to take out a "bigger" loan to expand his business further by making bacon and sausages. Because of his success, he was nominated as a representative of the small businessmen at ECLOF Kenya's quarterly meetings. In that capacity, he advocated (1) a lower interest rate and a more flexible way of repayment for those whose businesses do not pick up as fast as they imagined they would and (2) trust, maturity and understanding between group members as they co-guarantee one another.

Case 3: Grace Wanjiru – improving the household's income situation

Grace Wanjiru was a female household head living with her children in the Free Area region of Nakuru Municipality. As a source of livelihood for herself and her children, Grace started keeping *kienyeji* (local) chickens. And as a survival strategy, Grace joined an all-women "merry-go-round" group in the neighbourhood. The group members – mainly women of the same socio-economic status – contributed money on a weekly basis. The total weekly contribution was given to each member on a rotating basis. This was how Grace started and managed to sustain her "small poultry-keeping venture", as she called it.

Grace heard about ECLOF Kenya in 2002 from a friend and later shared the idea with the members of her merry-go-round group. Despite a lack of enthusiasm among some members, Grace and a few others managed to recruit some women outside their group and then registered as a self-help group. They did not, however, disband their merry-go-round group, which was still active at the time of the interview. Together, they attended ECLOF's training programme, which Grace described as "educative".

Grace, like many of her self-help group members, was given the Ksh. 20,000 she had applied for and that she used "to buy layers". She actually used half of the amount to buy 200 layers. The other half was used to buy chicken feed, as well as to erect a wooden enclosure for them. She had meanwhile re-paid her loan, although initially with difficulty. In addition to the profit she got from the sale of eggs, the money she received from the merry-go-round assisted her a great deal in repaying the loan. With the loan from ECLOF, Grace counted herself lucky as "a promising urban farmer who supplies eggs to various outlets in town".

At the time of the interview, Grace was collecting an average of three crates of eggs per day and selling them at Ksh. 150 per crate. Even though some of the layers died of disease, her monthly income had increased from less than Ksh. 3,000 to about Ksh. 10,000. Thanks to the loan, Grace had been able to (1) increase her chicken stock from fewer than 20 local chickens to over 100 layers, (2) put up a structure for them, (3) increase her monthly income threefold, (4) easily purchase feed for the layers, (5) take care of her children, and (6) improve her household's income situation and, in the long run, the family's standard of living.

Case 4: Jane Njeri – increasing milk production

Jane Njeri learned about ECLOF Kenya in 2003 during an outreach and promotion programme conducted by the credit officers in ECLOF's Nakuru office. Together with some like-minded friends and neighbours, they formed a self-help group as required by ECLOF. This group acted at the same time as a merry-go-round group, meeting weekly to discuss issues affecting them and presenting the total weekly contribution to one member on a rotating basis.

Jane, who already kept a cow for milk (zero-grazing), applied for a Ksh. 30,000 loan to buy another cow. However, after being granted the loan, she instead bought two heifers. The heifers matured into milk-producing cows and gave birth. Jane was proud that her milk production had more than tripled from about 5 litres to over 15 litres per day, raising her monthly income from Ksh. 3,000 to about Ksh. 10,000.

Jane said that before joining ECLOF, she could not raise the capital to buy another cow, while her husband, who was very supportive of this project, was not able to help her because of his meagre earnings. The profits from milk sales had gone a long way to improve the household's income and food security situation: they no longer struggled to put food on the table, to educate their children or to take care of their elderly parents at home. The husband was now able to concentrate fully on his business without regularly eating into his profits as before. Jane admitted that she initially had difficulties in repaying her loan but, with the support of her husband, she had managed. By the time of the interview, she had finished repaying it. In addition, she had planted Napier grass for her cows and had erected a good shelter for them using the profits from her milk sales.

In summary, this chapter showed that support, be it technical or financial, does make a difference for those involved. This applies particularly to urban livestock keepers, as very few crop cultivators appeared to have received any assistance at all. Livestock keepers who were supported by professional officers showed better rearing practices and better waste management. ARDP-supported farmers realized a higher income from their sales of livestock products than their non-supported neighbours. And those urban farmers participating in the ECLOF Kenya (micro credit) programme managed to raise their income levels considerably.

8

The environment

Co-author: Ernest O. Nyandwaro

Introduction

Urban farming and the urban environment are closely related. It is therefore surprising that only a few studies have been done on the impact of urban farming on the urban environment – mainly technical studies measuring pollution, erosion, etc. – due to agricultural practices in town. The use of chemical inputs for crop cultivation is usually seen as a threat to the urban environment. Studies in Tanzania (Foeken *et al.* 2004; Yachkaschi 1997) showed widespread use of chemicals, some highly toxic. In Nakuru too, quite a number of crop cultivators were found to use chemical inputs (see Chapter 4).

Another matter of concern is the use of untreated sewage water for irrigating crops. This appeared to be very common among spinach growers in Dar es Salaam (Mlozi 1999), although it did not necessarily affect the quality of the vegetables or soil (Muster 1997). Over time, the use of sewage water for irrigation can be harmful to the soil, as a crust consisting of particles sediment appears over the soil, causing an increase of compaction and making the soil more acidic. Eventually, some crops cannot grow anymore (Dennery 1995). In the Nakuru survey of 1999, only one crop cultivator was found to use sewage water for irrigation (see Chapter 4). That does not mean, however, that the practice was rare. On the contrary, in certain areas, for instance in the area known as Block 14 (just north of the old sewage treatment plant), untreated sewage water was widely used. How harmful this practice is was a matter of dispute in 2001, even leading to a court case (see Box 8.1).

Box 8.1: Safety of "sewage crops"

A dispute regarding the safety of crops irrigated with raw sewage water in Nakuru town reached the national press in an article entitled "Sewage crops' safety war may end up in court" that appeared in *Horizon*, a supplement to Kenya's biggest newspaper, the *Daily Nation* (8 November 2001). In June 2001, Ms Mary Muthoni threatened to go to court to get compensation for the loss of her *sukuma wiki*, after Nakuru Municipal Council had repeatedly raided her plot near the Kivumbini sewage treatment plant. According to the Municipal Chief Public Health Officer, Ms Muthoni's vegetables were the cause of a typhoid outbreak in May because of her use of sewage water to irrigate her crops. But Ms Muthoni claimed that her crops were not contaminated, a claim backed up by a laboratory report. Moreover, according to at least two microbiologists, getting typhoid from plants irrigated with sewage water was simply not possible. It was the same report that caused 13 other farmers, who admitted irrigating their gardens with sewage water, of going to court seeking an order preventing the Council from destroying their vegetables.

There was more agreement on the danger of tapeworms and roundworms that lay their eggs on vegetable leaves. Tapeworms thrive on human waste, which is an important component of the sewage water. The threat of infection is real, especially if the vegetables are eaten raw, as one microbiologist said. However, that can be prevented by properly washing vegetables before using them.

According to the Public Health Officer (PHO), another risk involved heavy metals, such as mercury, which has a cumulative effect when it gets into the human body. "The industrial waste water which enters into the Council's main sewage system contains mercury that is absorbed by the plants which are thus not suitable for human consumption", he said. In this, he was contradicted by the head of the Department of Environmental Health of the Ministry of Health, who said that "there is no evidence that the vegetables contain any significant levels of heavy metals", adding that "sewage water will rarely contain mercury". He did warn, however, that vegetables grown along major highways could absorb lead from car fumes.

Despite all this, the PHO's viewpoint was very clear. "The Council will continue destroying the vegetables until the farmers realise the dangers they pose to the public." He also called on members of the public to stop eating *sukuma wiki* irrigated with raw sewage water because of adverse effects on health. The Nakuru District Commissioner added to this that he would "mobilise local leaders to sensitise the public to the dangers of consuming *sukuma wiki*".

The soil in which a crop is being cultivated can be polluted by other sources as well, such as river water, industrial dumping or exhaust fumes. Along main roads and rivers in Dar es Salaam, fairly high heavy-metal concentrations have been recorded (Amend & Mwaisango 1998; Sawio 1996). Too intensive land use may exhaust soil fertility and even increase erosion, which was a matter of concern in Harare (Bowyer-Bower & Drakakis-Smith 1996; Bowyer-Bower 2002; Drakakis-Smith *et al.* 1995).

Livestock keeping in town can be harmful to the urban environment in a number of ways, as was for instance shown in some Tanzanian studies (Mlozi 1997, 1999; Mosha 1991). Firstly, livestock freely roaming around cause soil erosion and sometimes traffic accidents and can also destroy ornamental plants, lawns, water pipes, telephone lines, fences, etc. Secondly, domestic animals transmit diseases that can afflict humans and circulate among other animals. Thirdly, animal dung left to decompose in the compounds or along roads produces an odour (e.g. ammonia) and is a breeding ground for harmful bacteria and flies. Animal dung is also a source of tetanus. Slurry containing dung, urine and water, as seen in many compounds with cattle, chicken and pigs, attracts disease-causing vectors such as mosquitoes. Shauri (1988) found that the majority of a group of livestock keepers in Dar es Salaam dumped the dung along road verges. These dumping practices cause pollution of the ground and of drinking water. The situation in Nakuru was better (see Chapter 5), as 'only' a third of the livestock keepers dumped animal waste in the street. Most of this waste came from small livestock.

Crop cultivation and livestock keeping offer the possibility of recycling nutrients. For instance, among a group of crop cultivators in low-density areas in Dar es Salaam, 90% said they used organic matter in their gardens, be it chicken droppings, cattle manure or both (Mlozi 1998). In Nakuru, over half of the crop cultivators used manure as fertiliser.[1] The reverse, i.e. feeding livestock with crop residues, appeared to be common as well, particularly for large livestock.[2] On a larger scale, composting urban solid waste can serve both the urban environment and the production of crops in town. Extension workers could play an important role in the promotion of organic farming but, as Nkonya (1997) observed in Morogoro town, Tanzania, urban farmers do not receive adequate extension advice because the extension workers are not adequately trained in environmental issues. In Nakuru, however, there are indications that extension did have a positive impact on farming practices, as was shown in Chapter 7 (Table 7.1). For instance, those livestock keepers receiving (professional)

---

1 See Chapter 4, Table 4.2.

2 See Chapter 5, Table 5.4.

assistance were more inclined to feed their animals with crop residues and were less inclined to dump animal waste in the street.

The remainder of this chapter deals with three questions regarding the relationship between urban farming and the urban environment in Nakuru town. (1) To what extent are Nakuru farmers aware of the environmental impact of their activities? (2) What are the perceptions and attitudes among non-farmers and officials regarding farming in Nakuru town? (3) How polluted are the soil, water sources and crops?[3] The data were collected in the context of a separate study of NUAP that resulted in a Masters thesis (Nyandwaro, forthcoming).[4]

Urban farmers' environmental awareness

For the 'awareness' part of the study, 60 from the 209 urban farmers of the main survey (1999) were selected as a sub-sample. The 209 were divided into three groups – crop cultivators, livestock keepers and mixed farmers – and from each group 20 cases were randomly selected. As a result, the 60 cases included 40 crop cultivators and 40 livestock keepers.[5]

Almost 60% of the crop cultivators said they were aware of the pollution that could be caused by inputs (Table 8.1). Those aware of the problem mentioned soil pollution, water pollution, crop pollution or a combination of these. Asked whether they were planning to do something about it, over half of them said they would, indicating such measures as stopping the use of chemical inputs, using alternative inputs and/or avoiding the overuse of inputs.

Less than half of the livestock keepers were aware that their animals could be a menace to the neighbours or the urban environment (Table 8.1), mentioning bad smells, soil and/or water pollution, soil erosion, the transmission of disease, noise and traffic accidents. Various ways of dealing with these problems were mentioned, such as (in order of frequency) restraining animals from going into their neighbour's compound, seeking veterinary services to keep diseases at

3 It should be noted that the chapter does *not* deal with the impact of the urban environment on urban farming.

4 We gratefully acknowledge the funding of this study by Agropolis (IDRC).

5 In strictly methodological terms, this selection procedure may not be the most elegant one, but for the restricted purpose of this section it is acceptable. It was done because of the limited funds and time available in a Kenyan Masters study (consisting of a 'social' part – awareness, perceptions, attitudes – and a technical part, i.e. heavy metal concentrations). Some basic comparisons of the three groups of 20 are made in the Masters thesis (Nyandwaro forthcoming). In the present book, two of the groups are compared in the section on disposal of waste from livestock (Chapter 5).

bay, feeding their animals well to limit noise, ensuring proper waste disposal, cleaning the sheds daily, and reducing the number of livestock they had. Two respondents indicated that they wanted to start zero-grazing units as a means of reducing the menace their animals presented.

The large majority of livestock keepers were satisfied with the way they disposed of their animals' waste. Some were not, however. Asked what they intended to do about it, some planned to turn the waste into manure, while another wanted to use the waste for biogas production.

Table 8.1 Urban farmers' awareness of the environmental impact of their activities (% "yes")

(N=)	crop cultivators (40)	livestock keepers (40)
Are you aware of the pollution caused by crop inputs?	58	
Are you aware that livestock can be a menace?		43
Are you satisfied with the way you dispose of your animals' waste?		85
Do neighbours ever complain about your farming activities?	15	18
Is anyone giving you information about environmental pollution?	20	30

Source: Nyandwaro (forthcoming).

Both crop cultivators and livestock keepers were asked whether neighbours ever complained about their farming activities. As can be seen in Table 8.1, the majority of neighbours never complained (at least, according to the farmers). Those who had complained about crop cultivation did so because of chemicals ending up in their water sources. Complaints regarding livestock keeping referred to the destruction of crops and to the dirty conditions in compounds due to animal waste.

A minority of both crop cultivators and livestock keepers said that they had been or were being made aware of the environmental consequences of their activities by others (which seems to contrast with the above-mentioned indication that extension did have a positive impact on farming practices). Asked who those 'others' were, the livestock keepers mentioned veterinary officers, Municipal Council officers and also their own children telling them what they had learned at school about this particular issue. Their advice was to maintain general cleanliness (i.e. proper waste disposal) and keep large animals in zero-grazing. Crop cultivators mentioned agricultural officers, the electronic media and also "what was learned at school". Awareness creation consisted mainly of warnings about chemicals affecting crops and consumers; and information about environmental methods of disease control and the correct application of chemicals.

Perceptions and attitudes

Questions on perceptions and attitudes about relationships between urban farming and the urban environment in Nakuru town were put to a group of 20 non-farmers and 18 officials. The non-farmers were selected as the nearest non-farming neighbours of 10 crop cultivators and 10 livestock keepers. The officials consisted of ten municipal officers and eight representatives from other organizations, such as NGOs and CBOs.

According to most non-farmers, urban farming and urban pollution have been increasing in Nakuru town (Table 8.2). Interestingly, most officials had the opposite view. This may be due to the fact that officials tend to be transferred quite regularly, so they may have a much shorter time frame in mind than the non-farmers. In both groups, however, many think that farming in town causes pollution and in both groups "dirtying the environment" was mentioned in the first place, apparently referring to livestock keeping in particular. That was also the case with another common example, namely "the destruction of flower

Table 8.2 Urban non-farmers' and officials' perceptions regarding the environmental impact of farming in Nakuru town

		non-farmers (n=20)	officials (n=18)
▪ Has urban farming increased since you came here?	% yes	80	17
▪ Has pollution in Nakuru town increased?	% yes	70	22
▪ Does farming in town cause pollution?	% yes	?*	50
▪ Is farming in town a hygienic activity?	% no	70	67
▪ What is the quality of the crops cultivated in town compared with crops cultivated in rural areas?	% worse	70	61
▪ Are the crops growing at the dump polluted?	% yes	90	72
▪ Are crops grown near large industrial sites polluted?	% yes	50	61
▪ Are crops grown along roads contaminated?	% yes	50	72
▪ Do you know that some people use sewage water to irrigate crops?	% yes	100	89
▪ Do you think those crops are contaminated?	% yes	100	78

* Majority, but exact figure is not given in the source.
Source: Nyandwaro (forthcoming).

beds", to which some of the non-farmers added "bad smells" and some of the officials added "diseases" as well. A few complained about "increased crime due to maize" (as a hiding place for thugs), thereby referring to crop cultivation. A 'typical' problem for officials, also related to crop cultivation, was that farming in town "takes up space for urban development".

When asked how the negative effects of farming in town could be tackled, it was interesting to see that the non-farmers pointed solely to the government that should, for instance, "impose strict laws regarding waste disposal", "reduce the numbers of farmers", "ban farming in town altogether", and/or "support farmers in improving hygiene". To some extent, the officials also pointed at themselves, for instance by stating that "veterinary services should be provided", "waste should be collected regularly by the Municipal Council" and/or "keeping animals near residential areas should not be allowed". On the other hand, they also saw a responsibility on the farmers' side and for livestock keepers in particular. Some of the officials stated that they "should clean the animal sheds regularly", thereby suggesting that standards of cleanliness were often not up to standard.

Two-thirds of both non-farmers and officials perceived farming in town as an unhygienic activity (Table 8.2). It is not surprising, then, that they also perceived the quality of the crops cultivated in town as being lower than those from the rural areas. Crops cultivated at certain locations – the dump, near large industrial companies, along roads – are perceived by many as being polluted. This applies particularly to the dump (which is confirmed by the measurements presented in the next section). All non-farmers and most (but not all!) of the officials indicated that they knew about the use of sewage water for irrigating crops in certain areas. There was a common view regarding the contamination of such crops, although apparently some of the officials had a different view on that. Since the fieldwork took place before the court case presented in Box 8.1, it is possible that these officials have since changed their minds.

To assess the attitudes of non-farming neighbours and officials regarding farming in town, they were asked to give their opinions on a number of statements on cultivating crops, the keeping of large livestock and the keeping of small livestock in Nakuru town. The results are presented in Table 8.3. Note that the statements are either unfavourable, indicated by (-), or favourable, indicated by (+). Moreover, the statements are presented in a rough sequence from 'most unfavourable' to 'most favourable'. Note also that the absolute numbers in the two groups are small and the percentages should therefore at best be seen as indications.

A number of conclusions can be drawn from the table. First, there was a fairly general conviction among both non-farmers and officials that urban farming is important for the urban food supply. However at the same time, farming in its present form was considered negatively by many because "it is bad for the environment". This applied in particular to growing crops and keeping large livestock. Attitudes regarding the keeping of small livestock in relation to the urban environment were somewhat more favourable among both groups.

Second, opinions became more favourable if farming was better kept under control. As for crop cultivation, the majority of the respondents in both groups agreed with the statement "acceptable in designated areas only". Regarding keeping large livestock, it is interesting to see that whereas in the first instance almost a half of the officials agreed with the strong statement that it "should be forbidden", a few statements later (literally), quite a number of them apparently

Table 8.3 Urban non-farmers' and officials' attitudes regarding urban farming (%)

	20 non-farmers		18 officials	
	agree	dis-agree	agree	dis-agree
Growing crops in town …				
… should be forbidden	20	35	22	33
… is a bad thing	65	10	28	61
… is bad for the environment	70	10	67	11
… should be allowed in designated areas only	70	10	78	0
… is important for the urban food supply	80	10	61	28
… contributes to a better environment in town	0	100	0	94
… can solve urban environmental problems	10	30	0	67
… should be stimulated by the government	35	30	33	17
Keeping large livestock in town …				
… should be forbidden	20	30	47	22
… is a bad thing	20	10	33	22
… is bad for the environment	60	10	56	22
… is acceptable if controlled by the government	60	20	11	78
… is acceptable if waste disposal is controlled	90	0	78	6
… is important for the urban food supply	80	5	89	6
… should be stimulated by the government	55	10	47	11
Keeping small livestock in town …				
… should be forbidden	35	20	11	44
… is a bad thing	25	35	11	44
… is bad for the environment	35	15	39	22
… is acceptable if controlled by the government	50	20	61	22
… is acceptable if waste disposal is controlled	65	20	33	44
… is important for the urban food supply	80	10	78	6
… should be stimulated by the government	45	35	61	11

Notes: *Agree* includes 'strongly agree' and 'agree'.
Disagree includes 'strongly disagree' and 'disagree'.
The answer 'no yes, no no' is not included in the table; hence, the totals do not add up to 100%.
Source: Nyandwaro (forthcoming).

changed their minds ("should be accepted if waste disposal is controlled"). All the non-farmers but one held the same opinion. This is an indication of the seriousness of the problem of waste from livestock in town. More generally, it also shows how much farming in town is related to the urban environment in the people's minds. Very few, however, thought that crop cultivation is or might be beneficial to the urban environment. All respondents except one disagreed with the statement that crop cultivation "contributes to a better environment in town". Moreover, only two (out of 38) respondents thought that growing crops "can solve urban environmental problems", though it should be noted that the majority of the non-farmers and a third of the officials had no opinion about this.

Third, the non-farmers and the officials appeared to have quite similar attitudes regarding urban farming in Nakuru town. One difference concerns the general attitude to crop cultivation in town. Whereas most non-farmers considered this as "a bad thing", the majority of the officials did not agree, even though most felt that it was bad for the urban environment. Another difference has to do with the role of the government in relation to the keeping of large livestock. Many non-farmers saw a role for the government ("acceptable if controlled by the government") but the officials did not share that opinion. Although this is in line with the findings regarding the perceptions presented above, it is surprising that many officials saw a controlling role for the government concerning small livestock. Like the non-farmers, about half of them agreed with the statement that keeping livestock, either large or small, "should be stimulated by the government". Very few respondents actually disagreed with this. It is also interesting to see that this promoting and encouraging role by the authorities was more popular regarding livestock keeping than crop cultivation. Although we have no ready explanation for this, it might be related to the fact that rearing livestock is a more commercial activity compared to crop cultivation.

Heavy metal concentrations

To ascertain the extent to which the environment in Nakuru town was polluted in areas where urban farming was taking place, heavy metal concentrations were measured in soils, water used for irrigation and crops grown. Specific variables were distance from the main road, age of the crops grown, and site. Samples of soil, water – both sewage water and tap water – and crops were collected and analysed in a laboratory for four types of heavy metals:, namely zinc (Zn), lead (Pb), cadmium (Cd) and mercury (Hg). Twelve sampling sites were selected, in such a way that together they could be considered representative of the whole of Nakuru town. Most of them were located in the 'clusters' used for the general survey. Specific sites included the dump in the northwest of the town (where there is a lot of farming), areas near roadsides and places where sewage water was used for irrigation. A brief characterization of the twelve sites is presented in Table 8.4. Site 6 (Lanet) was considered the 'control area' and was thought to be unpolluted.[6]

Table 8.4 Characteristics of the sampling sites

Sampling site	Observable characteristics
1 Mwariki	Sloping plot near the sewage plant but no irrigation water used for farming.
2 Rhonda Sewage	Near the sewage plant, with extensive use of sewage water.
3 London (dump)	The dump with a lot of non-biodegradable, synthetic materials. A large part of the site is covered with a thin layer of soil for crop-cultivation purposes.
4 Lanet Free Area	Use of tap water for irrigation.
5 Rhonda Pondamali	Near the market place, very sandy and neglected farming area. Plots were not weeded.
6 Lanet	Relatively big farm with no water for irrigation and located a long way from the urban set-up.
7 Kivumbini	Occasional use of sewage water for irrigation, especially during the dry season.
8 Rhonda Kaptembwa	Extensive use of organic manure from the animals kept on the farm.
9 Kabachia	Area next to the road with backyard farming, plots covered with ashes of burnt tins and tyres.
10 Rhonda Muslim	Area with high use of organic manure but regularly cleaned.
11 Kaloleni	Next to the sewage plant, with extensive use of sewage water.
12 Milimani	Staff residential area with backyard farming.

Source: Nyandwaro (forthcoming).

Soils

The soils at all the sampling sites consist of haphic phaezems. The area is tertiary volcanic with flood plains. It has flat to gently undulating plough ridges that are good for agricultural activities. There is a considerable percentage of gravel, silt and clay in the soil, which may be responsible for an increased capacity of adsorption of heavy metals from the topsoil. Such finer textured soils are less permeable, so plants have a longer time to absorb trace elements. Soil-profile

6 See Nyandwaro (forthcoming) for a description of the research methods, the laboratory analysis and a detailed description of the sampling sites.

analysis showed that the water level is deep, which indicates that the area exposed to water is low with slow leaching water velocity.

Annex 8, Table A8.1 summarizes the measured concentration levels of heavy metals in the soil, while Table 8.5 presents the sampling sites with the highest concentrations. Relatively high levels of heavy metals were found at three sampling sites, namely Rhonda Sewage, London (dump), and to a lesser extent Kaloleni. In Rhonda Sewage, zinc (Zn) levels in particular were substantially higher than the limits allowed by the WHO (50-150mg/l). Compared with other sites, levels of most elements were also found to be relatively high at Rhonda Sewage and the dump. These high levels can likely be attributed to the use of sewage water for irrigation (Rhonda Sewage, Kaloleni) and the presence of synthetic and domestic solid wastes at the dump.

Table 8.5 Sites with relatively high heavy metal concentrations in the soils (mg/l)

Sampling site↓		Zn	Pb	Cd	Hg
	WHO standards→	50-150	5-10	0.05-0.2	<1
Rhonda Sewage		241.0	4.5	0.21	0.10
London (dump)		135.6	4.5	0.51	0.61
Kaloleni		53.5	1.4	0.08	0.14

Source: Annex 8, Table A8.1.

As for the relationship between the concentration levels, on the one hand, and distance from the main road, on the other, zinc (Zn) and lead (Pb) levels were found to be higher at sites nearer to the roadside. But for cadmium (Cd) and mercury (Hg), it was the other way around, namely levels were higher further away from the roadside. However, the relationship was statistically not significant ($p>0.05$) and the sample was too small to permit any generalization.

Water

Table 8.6 shows the measured concentration levels of the four trace elements as found in tap water and in sewage water, both used for irrigation of the cultivated

Table 8.6 Heavy metal concentrations in water, by type of water source (ppm*)

Water source	Zn	Pb	Cd	Hg
Tap water	0.13	0.09	0.08	0.08
Sewage water	0.24	0.11	0.90	0.11

* Parts per million.

Source: Nyandwaro (forthcoming).

crops. For all four heavy metals, cadmium in particular, concentration levels were found to be higher in sewage water than in tap water.

Crops

It was decided to use African spinach (*amaranthus* spp) to measure heavy metal concentrations in crops. This was for three reasons: first, it was available at *all* sampling sites all year round; second, it tolerates environmental stress so the age of the plant could be used as a variable; and third, the crop is popular in the Nakuru households and consumed regularly.

Table A8.2 in Annex 8 shows the heavy metal concentrations in the leaves of this crop at the twelve sampling sites, while Table 8.7 shows the sites with the highest concentrations. Again, Rhonda Sewage and the dump stand out because of the comparatively high levels of zinc (Zn). In Kaloleni, however, where sewage water is used too, zinc levels were relatively low. This is likely to be due to the fact that, compared to Rhonda Sewage, irrigation was less intensive and the area has a steeper gradient.

Table 8.7 Sites with relatively high heavy metal concentrations in Amaranthus (mg/l)

Sampling site↓		Zn	Pb	Cd	Hg
	WHO standards→	50-150	5-10	0.05-0.2	<1
Rhonda Sewage		210.0	10.0	0.10	0.21
London (dump)		130.0	4.6	0.09	0.50
Mwariki		26.8	10.3	0.02	0.10
Kabachia		4.7	7.8	0.11	0.09

Source: Annex 8, Table A8.1.

Concentration levels of lead (Pb) were comparatively high in Mwariki, Rhonda Sewage and Kabachia. In Mwariki, this is mainly because of its proximity to the sewage plant, and in Rhonda Sewage (again) to the intensive use of sewage water for irrigation. The high levels of lead in Kabachia can be explained by its proximity to the road and by the fact that shortly before the plant samples were collected, some tires had been burnt there. For the same reasons, cadmium (Cd) levels were comparatively high at the same sampling sites (Rhonda Sewage and Kabachia). In Rhonda Muslim, a comparatively high cadmium level was found as well due to the use of raw industrial effluents from the industrial area for irrigation, in combination with its proximity to the road and the use of inorganic manure. The cadmium levels in the plants cultivated in areas where sewage water was used for irrigation were much lower than the cadmium levels in the sewage water itself. This is due to a high degree of leaching of the soils and because the uptake levels of the metal from the soils is 'shared' by the crops and the weeds on the farmland. Finally, levels of mercury (Hg) were generally low. The only exception was the dump, although there the level still appeared to be below the WHO's permitted limits.

As for the distance to the main roads, apart from zinc, concentrations of heavy metals were found to be higher in plants closer to the roads. In terms of the age of the plants, younger plants appeared to contain (statistically significant: $p<0.05$) higher levels of zinc, cadmium and mercury than mature plants. With lead, it appeared to be the other way around, due to long-term accumulations, although this relationship was statistically not significant ($p=0.0990$).

In summary, the chapter has shown that quite a number of urban farmers in Nakuru were not aware that their activities could have a negative impact on the urban environment. Therefore, it is not surprising that the farmers' neighbours and officials generally held the view that farming in town is bad for the environment. Their view became more favourable, however, if farming in town was better controlled. An important aspect in this regard concerns the comparatively high levels of heavy metals in soils and plants irrigated with sewage water, in plants cultivated at the dump, and in plants growing close to roads. This indicates that the Public Health Officer (see Box 8.1) did have a point about the heavy metals in the raw sewage water used to irrigate crops. Yet, it must also be noted that the present study found that the concentrations at these sites were high *in comparison with* other sites, which is not the same as saying that they pose a serious health threat for the people consuming these plants (although some concentrations were indeed higher than the limits set by the WHO). Another conclusion is that the Council should not only focus on areas where sewage water is used but on the dump as well.

Photo 11 Nakuru's dump. Unseparated waste on the left, maize cultivation on earth-covered waste on the right. (Dick Foeken, 1999)

Photo 12 Dairy cattle roaming in the street feeding on household waste in Stadium area. (Sam Owuor, 2004)

9

The schools

Co-author: Correta E. Odera

Institutional farming

Besides farming by individual households, farming by institutions is also common in African towns and cities. However, farming by urban institutions has been largely overlooked or neglected by both researchers and policy-makers. Besides state farms located within the boundaries of urban centres, institutions that perform agriculture in town include monasteries and convents, factories, prisons, schools, hospitals and the army. For instance, a large factory in Morogoro, Tanzania (Tanzania Tobacco Processors Ltd.) exploits vegetable and fruit fields, partly for consumption in the firm's canteen and partly to sell to individual buyers.[1] In East Africa, most prisons have farms, an example being Nakuru Prison Farm (see Box 9.1).

Probably the most important type of institutional urban agriculture is school farming. Data available from Dar es Salaam are based on an aerial survey carried out in 1999 (Dongus 2000) when 33 schools were engaged in vegetable production. Of these, 30 were primary schools and the others were secondary schools. The majority of the schools (23) practised rain-fed farming, the others irrigated their crops. A total of about 62 acres were under cultivation, i.e. an average of 1.9 acres per school. Comparing these data with those from aerial maps of 1992 showed that six of the 33 schools did not grow vegetables in 1992, but started doing so afterwards. However, another 16 schools stopped

---

1 Personal communication at the site, 1999.

Box 9.1: Nakuru Prison Farm

All prison departments in every province of Kenya farms. Nakuru Prison Farm covers 1000 acres, with citrus trees being grown on 10 acres, 300 acres is for pasture and the rest is for cultivation. The crops grown include maize, beans, peas and a wide range of vegetables (spinach, green peppers, egg plants, cabbages, carrots, sweet potatoes, sugar cane, and a variety of local vegetables such as *kunde, managu, terere* and *saget*). Horticultural crops include oranges, lemons, tangerines, grape fruits, limes, avocadoes, mangoes, custard apple, passion fruits, paw paws, tomatoes, apples, pears, plums, water melons, apricots and bananas. The livestock kept include cows, goats, sheep, chicken and bees. Goats and sheep are reared in an open free-range system in and around the prison. In September 2000 there were 80 cows, 70 goats, 45 sheep and 100 local (*kienyeji*) chickens.

All the produce – vegetables, fruit, seedlings, milk, eggs, animals and honey – is sold. The prison authorities buy food from the farm at subsidised rates to indirectly generate income to help covering operational costs. The rest is sold to the prison's staff (at low rates) and to people from outside (at commercial rates). Inmates working in the fields can only benefit through eating overripe fruits. Some inmates, such as young children who are living with their imprisoned mothers, are supplied with milk, albeit only after the recommendation from the prison's doctor.

The prison inmates carry out the farming activities, not necessarily as a punishment but to learn basic farming skills. It is a means of rehabilitation and keeping the inmates busy. There are even stories of former inmates who have become farmers after this kind of training.

The prison's farm is like a farm demonstration centre where anybody is welcome and advice is given on various aspects of farming. Sometimes a minimum fee is charged depending on the nature of the consultation. On-plot demonstrations as well as on-plot scientific research on various types of breeds of crops and fruit trees are also done (an example being research into the cultivation of pears in a tropical climate like that in Nakuru). Most of the results of these research activities are disseminated through the Agricultural Society of Kenya's annual shows.

Source: Mureithi 2000.

cultivating between 1992 and 1999, mostly because of a shortage of rains, new school buildings, the planting of trees and/or flowers, or a combination of these reasons.

We know of only one in-depth study on school farming that was carried out in city of Cagayan de Oro in the Philippines (Potutan *et al.* 1999). Nearly all public primary schools there have gardens, which are tilled by the pupils (assisted by their mothers) under the supervision of teachers. The study in Cagayan de Oro reported that the activity was beneficial in various ways, contributing to the children's household economy and food security, the production of cheap and nutritious foods, the consumption of healthy food, the urban environment, and communalism and cooperation in the local community.

School farming in Kenya

School farming is normal in Kenya, particularly in the rural areas where it dates back to the colonial period. Since 1988, practical subjects like agriculture are on the curriculum at secondary school and for those students who choose agriculture as a practical subject, farming is compulsory. It is, therefore, to be in all (rural) secondary schools in Kenya (Mwago 2000).

Farming in urban schools started mainly for aesthetic reasons, with the planting of flowers and trees. There was no need in the past to grow food because the government subsidised school feeding programmes and the pupils needed to pay very little. But in 1978 government food subsidies stopped and schools started to feel the need to grow food crops instead of flowers and trees. Some schools had already started to grow crops, for example the first school in Nakuru to do so was Bahati Secondary School, which started in 1970 (Mwago 2000).

Farming in primary schools was promoted by the government through the so-called 4-K clubs, an acronym for *Kuungana, Kufanya, Kusaidia, Kenya*, which means "get together, act and help Kenya". The major goals of the programme were: (1) to teach the youth improved methods of agriculture; (2) to teach the youth to appreciate agriculture and the dignity of labour with respect for agriculture as a profession; (3) to help the youth produce food for their families and to sell; (4) to develop leadership skills among the youth and adults through voluntary participation in agricultural programmes; and (5) to change adult farmers' attitudes and practices (Odera forthcoming).

Today, the farming activities in schools are carried out by either these 4-K clubs or by Young Farmers' Clubs, the agriculture class or the school itself. The 4-K clubs are supervised by a teacher, usually the agriculture teacher, who also acts as patron of the club. However, most decisions are taken by the club members themselves, including the decision about which crop(s) to grow and what will be done with the produce. In schools with an agriculture class, the choice of crops is determined by the syllabus and farming is a practical session of what has been taught in class. Each student has his/her individual plot and the crops they grow are assessed for examination purposes (Odera forthcoming).

Decisions about production depend on the way farming is organised, on the type of school, and on the persons in charge. Pupils organized in 4-K clubs have a say in the way the produce is used: crops may be sold or taken home for consumption. In boarding schools, the produce is often used for school meals.

In schools with a large area of land, the administration is more likely to have a say in the crop's destination because larger sums of money may be involved (Mwago 2000).

School farming in Nakuru town[2]

In 2000, there were about 80 primary and secondary schools in Nakuru town. The schools can be distinguished in three different ways: by category (primary/-secondary), by type (day/boarding) and by management (government/private). The majority (71%) were primary schools. Another 26% were secondary schools, while there were two schools with both a primary and a secondary department. Most schools (81%) were day schools. The others were either boarding schools or a mixture with day pupils and boarders. Finally, most schools (75%) were government schools. The other 25% were run by religious institutions, mainly churches.

Frequency of farming
Fifty of the 80 schools in Nakuru town appeared to perform some kind of farming activity (Table 9.1). Farming is much more common among secondary schools (90%) than among primary schools. The higher percentage of farming among boarding schools is related to the fact that most boarding schools are secondary schools. Finally, it makes no difference whether a school is run by the government or a private institution as to whether farming takes place.

Among the 30 schools without any farming activity, several reasons were mentioned for them not doing so, such as a lack of land, destruction of crops by animals (both livestock and wild animals from the nearby Lake Nakuru National Park), lack of rain, and "new school" (i.e. no time to set up a farm). However, most of the non-farming schools were aware of the farming activities of the other schools and of these, the majority showed at least an interest in starting some kind of agricultural activity.

Three-quarters (74%) of the farming schools only cultivated crops. Another 22% grew crops and kept livestock, while the remaining two schools only kept livestock. Hence, the overwhelming majority (48) of the 50 farming schools

---

[2] The findings presented in this section are based on data collected in September-October 2000 by Correta Odera in the context of a study for her MSc thesis (Odera forthcoming). What is presented in this chapter is based on a general survey among (almost all) 80 primary and secondary schools in Nakuru town. Respondents were either somebody from the school management (usually the headmaster) or the agriculture teacher or both.

Table 9.1 Frequency of school farming by school characteristics (%)

		(N=)	absolute	%
all schools		(80)	50	62.5
school category*	primary	(57)	30	52.6
	secondary	(21)	19	90.5
school type	day	(65)	38	58.5
	boarding/mixed	(15)	12	80.0
school management	government	(60)	37	61.7
	private	(20)	13	65.0

* The two schools with both primary and secondary departments have been omitted here. One appeared to farm.
Source: School survey 2000.

cultivated crops, while 13 kept livestock. The popularity of crops over livestock may partly be explained by the fact that growing crops is not only cheaper but also easier than keeping animals, for instance in terms of disease management and feeding. Considering that the students do most of the farming labour, the day-to-day care of animals can be problematic, especially during holiday periods. Finally, the government, in assessing agricultural practical work, has always laid more emphasis on crop production than on rearing animals.

Table 9.2 shows the persons responsible for farming in school. In most schools – and particularly in secondary schools – the students of the agriculture class are the ones who do the farming, for educational purposes. In about one third of the schools, the school management is either wholly responsible for the farming activities or does so with the pupils. Such 'combined' responsibility is more common in primary than in secondary schools. In only four schools were

Table 9.2 Persons responsible for farming by school category (%)*

	primary schools (N=30)	secondary schools (N=19)	all schools (N=50)
school management	40.0	21.1	32.0
Young Farmers' Club	3.3	15.8	8.0
4-K Club	26.7	-.-	18.0
agriculture class	43.3	84.2	58.0
students	6.7	5.3	6.0

* Totals exceed 100% due to combinations.
Source: School survey 2000.

there Young Farmers Clubs, while in nine primary schools farming was done as 4-K Club projects.

Few schools (seven) received any form of external support for their farming activities, either from an extension officer employed by the government or from an NGO. The types of support included training, demonstrations, financial assistance, equipment and veterinary services. All schools that received external support were primary schools. Why none of the secondary schools reported receiving support is not clear.

Crop cultivation

The average school plot for crop cultivation was 1.7 acres.[3] On these plots, thirteen different types of crops were cultivated in 1999 and 2000. The most common were kale, cabbage, spinach, beans, maize and Irish potatoes. Among the less important crops were cowpeas, carrots, onions and tomatoes. In general, crops cultivated by schools did not differ from those cultivated by individual households. However, harvests were very bad in both years, due to insufficient rainfall.

Farming techniques were generally very simple. The *jembe* (hoe) and *panga* (cutlass) were the most common tools used for crop cultivation. In only six schools was a tractor used as well.

The types of inputs used are shown in Table 9.3. The use of inorganic fertilisers and improved seeds and seedlings was widespread among the Nakuru schools, especially the secondary schools. This shows the seriousness with which the schools undertook farming activities. Six schools – all primary schools – did not use any inputs, not even fertilizer. This might be due to the relatively high costs involved. In general, chemical inputs were very commonly used in secondary-school crop cultivation. More traditional and more environmentally friendly inputs such as manure were used in a minority of the primary schools only. Crop residues were used by only one school and urban waste not at all.

Despite the unpredictable climate, irrigation was not very common (Table 9.3) and almost all schools depended largely on rainfall. Although several schools – especially secondary schools – also used water obtained "from the municipal council" (i.e. tap water), this was not sufficient. In general, water is quite scarce in Nakuru. Not surprisingly, therefore, all schools mentioned "lack of rain" as *the* major constraint.

---

3 Two extreme values are excluded here – one of 25 acres and one of 192 acres. The latter plot belongs to Nakuru High School and is solely used for growing Napier grass for the 60 head of cattle the school owns.

Table 9.3 Inputs used for crop cultivation by school category (%)

	primary schools (N=28)	secondary schools (N=18)	all schools (N=47)
no inputs used	21.4	-.-	12.8
farmyard manure	42.9	5.6	27.7
composite manure	21.4	5.6	17.0
chemical fertiliser	42.9	100.0	66.0
chemical pesticides	21.4	77.8	44.7
chemical insecticides	25.0	50.0	34.0
improved seeds/seedlings	39.3	61.1	48.9
irrigation	21.4	33.3	27.7

Source: School survey 2000.

Most of the labour needed for land preparation, planting, weeding, watering and harvesting was provided by the students. In some schools, however, others from the school community, like teachers and school workers, also gave a hand. In eleven schools, additional labour was hired, particularly for the heavy work of land preparation. Finally, one school located in a high-income area not far from the prison had the privilege of free labour by prison inmates. With the major contribution of labour for crop cultivation coming from the students, farming can be made a fairly cheap venture for schools. Moreover, students are supposed to gain in educational terms as well as acquiring practical skills that they can use outside school or apply later in life.

Livestock keeping

As mentioned above, livestock keeping was not common among the schools in Nakuru, as only 13 schools practised this kind of farming. Seven of these schools were secondary schools, five were primary schools and one was a school with both primary and secondary departments. This means that a third of the secondary schools in Nakuru kept some livestock, against less than 10% of the primary schools.

Cattle was the most common type of animal (in eight schools), followed by rabbits (five schools), chickens (four schools) and sheep and goats (both one school). The cows were kept for their milk, which was used for making tea at 10 am and at 4 pm for both students and teachers. The rabbits were used in the students' meals and the chickens in teachers' meals. Small livestock was also used for special school activities, for instance in biology lessons.

Except for the earlier mentioned Nakuru High School, the number of cattle in 2000 was modest, ranging from two to seven. The number of rabbits ranged from four to forty and chickens from two to twenty. The majority of the nine livestock-keeping schools kept only one type of animal; three schools had two types. One primary school in Kimathi Estate was exceptional in having two cows, one goat, 15 rabbits and five chickens. The numbers of animals change every year due to births, sales, deaths and thefts. This applies particularly to small livestock like rabbits and chickens. For instance, in the five schools with rabbits some 50 animals were sold in 1999, while 29 died the following year. The total number of chickens in 2000 was much lower than the previous year because about 100 were stolen.

The most common rearing systems were zero-grazing for the cattle and caging for the rabbits and chickens. Only one school had sufficient space in the compound to let their cattle graze.

Grass was the most common type of fodder for the animals, mainly for cattle. This was either obtained free of charge or bought at particular points in town (partly due to the drought). In some schools, cattle were (also) fed with Napier grass, which was grown by the school itself. The use of crop residues was practised in five schools and urban waste in three schools, mainly for rabbits and chickens. Rabbits were also fed with weeds from the school garden (two schools) and with kitchen leftovers (one school).

In terms of 'modern' inputs, cattle were better looked after than other types of livestock. For instance, veterinary drugs were given to the animals in eight schools, but only in those where there were cattle. In four schools, artificial insemination was done (though not in Nakuru High School where natural breeding was practised). Feed supplements for cattle were given in two schools.

In six schools, students were responsible for taking care of the animals, usually the agriculture class (five schools). In one school, the 4-K Club looked after the animals. All these cases concerned rabbits. Cattle and, to a lesser extent, chickens were taken care of by either school workers (five schools) or hired help (two schools). In one secondary school, however, the agriculture class was involved in looking after the five cows as well.

Destination of the produce

The produce (crops, milk and sometimes animals) from the school farms had three destinations: (a) the schools' feeding programme, (b) it was sold, and/or (c) it was taken home by pupils. As shown in Table 9.4, the school's feeding programme was the most common destination. This applied especially in

Table 9.4 Destination of farm produce by school category (%)*

	primary schools (N=25)	secondary schools (N=17)	all schools (N=43**)
- school feeding programme	44.0	88.2	60.5
- sold	52.0	23.5	41.9
- taken home	24.0	17.6	20.9

* Totals exceed 100% due to combinations.
** Includes one school with both primary and secondary education. For seven of the fifty farming schools, the destination of the produce was unknown.
Source: School survey 2000.

secondary schools.[4] By producing some of the food they needed for the feeding programme, a school saved money with which it could buy what the school did not produce itself.

Selling produce was more common in primary schools than in secondary schools. Produce could be sold to students and teachers (usually at reduced prices) or to the public. The way the money earned from produce sales was spent depended on who the seller was: the school or the students involved. If the school sold the produce, the extra income was mostly used to meet some of the school's financial obligations, such as paying bills for water and electricity and buying stationary. This could have been due to reductions in monetary support from the government. The government provides teachers and pays them but expects the schools to raise their own funds to run the schools (cost sharing). The schools get this money from the fees paid by students but it may not be enough, so selling farm produce is a way to raise the school's income. If it was left to the students to decide what was sold, they could normally decide how to use the income. The common choices were to organise a party for club members and/or to re-invest money in the next farming project.

In nine of the fifty farming schools, all or part of the produce was taken home by either students (eight schools) or teachers (one school). This occurred only when students were involved in farming. Whenever the school itself was involved, the produce was either used in the school's feeding programme or sold. It was therefore notable that at the one (primary) school where the teachers took the produce home with them, it was the agriculture class who did the farming (although some of the teachers did some of the work as well).

4 The same differences were found for school type (day versus boarding) and school management (government versus private): in nine of the ten farming boarding schools and in ten of the eleven farming private schools, farm produce was intended for feeding programmes.

School feeding programmes

In 1967, a school feeding programme was launched in Kenya by the National School Feeding Council of Kenya (NSFCK). By 1986 it had reached its peak, covering 60,000 pre-primary and primary-school children in 15 districts. Mainly due to financial constraints, the programme had to reduce its activities and by 1996 only 13,000 children were being reached in four districts, Nakuru District not being one of them. Currently, no subsidized school feeding programmes exist in Nakuru town. The administration of those feeding programmes that do exist rests with the head teacher or with private caterers who bring previously cooked food to the schools. Two studies, one in Kirinyaga District (Pieters *et al.* 1977) and one in Nyambene District (Meme 1996), showed that children participating in the NFSCK programme were better off in terms of nutritional status and school performance than those children not in the programme.

Over 60% of the schools in Nakuru had a school feeding programme in 2000 (Table 9.5). Feeding programmes appeared to be particularly common among secondary schools, boarding schools (obviously) and private schools. Feeding programmes were also more common in schools that farmed. Although not all farming schools had a feeding programme and not all schools with a feeding programme farmed, there was a significant positive correlation (p=.000) between the two. Finally, all schools that kept livestock appeared to have a feeding programme too, suggesting that animals were kept primarily because of the feeding programme (milk and meat).

Table 9.5 Frequency of school feeding programme by school characteristics (%)

		(N=)	absolute	%
all schools		(80)	50	62.5
school category*	primary	(57)	29	50.9
	secondary	(21)	20	95.2
school type	day	(65)	35	53.8
	boarding/mixed	(15)	15	100.0
school management	government	(60)	32	53.3
	private	(20)	18	90.0
school farming?	yes	(50)	39	78.0
	no	(30)	11	36.7
type of farming	crops	(48)	37	77.1
	livestock	(13)	13	100.0

* The two schools offering both primary and secondary education have been omitted here.
Source: School survey 2000.

Despite the fact that school feeding programmes have a relatively long history in Kenya, those in Nakuru town at the time of the study were mostly more recent. In over 70% of the schools with a feeding programme, the programme had started in 1995 or later. In almost half of the schools, the programme started in one of the three years preceding the survey and in almost a quarter of the schools even in the same year (2000). In only six schools had the programme been in operation before 1990. Although the reason behind this recent start of the feeding programmes in most schools was not requested, it may indicate that schools are feeling increasingly responsible for the well-being of their pupils in these times of economic stress. Children perform better at school with a decent lunch in their stomachs. Moreover, in schools where children have afternoon classes, it is often considered better to keep them at school at lunchtime to have some control over the time spent over lunch. Serving lunch also acts as a time-saving strategy for the school.

The meals offered by schools could be breakfast, morning-break tea, lunch, tea or supper. In three schools, all these meals were provided. In fourteen boarding schools, the three main meals (breakfast, lunch and supper) were served. Lunch was by far the most common meal served in schools.

Schools also differed regarding the persons eligible for the feeding programme. With the exception of one school, students always belonged to the target population but in some schools only the students in examination classes (standard 8 or form 4) were eligible, usually for lunch, to encourage them to spend the whole day at school to use their time to the best possible advantage. This was done in a bid to improve school performance (i.e. raise the average grade of the examination class). The charge students had to pay was usually included in their school fees as many parents failed to pay if it was charged separately. In half of the schools, school staff – be it teachers, other staff or all of them – could also benefit from the meals provided. Finally, in seven schools the feeding programme was only open for those students or staff "willing to pay for it". In these cases, it was usually a business venture organized by either a private caterer or a school teacher who cooked the food at home, brought it to school and sold it. Where this was the practice, the parents gave the children money for lunch and did not pay anything directly to the school.

The main source of the food served in the context of the programme was the local market. Forty-three per cent of the schools depended solely on purchased food and another 55% partly. For half of the schools (24)[5], their own farm pro-

---

5 This is fewer than the 26 schools where the destination of the school's farm produce was said to be the school feeding programme (see Table 9.4). It is possible that in answering the question about the source of food for the feeding programme, the

duce was one of the sources of the food, but never more than an additional source. A few schools had also received donations for this purpose.

It would be interesting to know whether schools that farmed and had a feeding programme as well put more energy into their farming activities than schools that also farmed but did *not* have a feeding programme. This question is difficult to answer, partly because no data had been collected specifically for this purpose, and partly because the second group (farming yes, feeding programme no) consisted of only eleven cases. However, given these limitations, on such variables as plot size, type of inputs, number of inputs and water source, no differences could be seen between the two groups. Hence, based on the data available, the question cannot be confirmed.

Perceived benefits of school farming

Schools can benefit from farming in several ways. The most frequently mentioned benefit (42%) was the money saved and that was spent on buying food for the school feeding programme. The second most frequently mentioned benefit (32%) was farming as a source of income, to help alleviate the financial burden placed on schools. A benefit of a very different kind, which was mentioned by a quarter of the respondents, was farming as a method of teaching agriculture as a subject. On the other hand, there were three respondents who did not understand how schools were gaining from farming in any way.

Students can also benefit from school farming in different ways. Over three-quarters of the respondents saw the learning experience as the major benefit for the students involved. About half of the respondents mentioned the source of food for them, either provided free or sold to them cheaply. Another benefit for at least some students was that by working at the school farm they were able to pay their school fees indirectly. Students who had problems paying school fees were used to doing all kinds of odd jobs on the farm, such as weeding. The money that would otherwise have been paid to hired staff was credited to the students' school account allowing him/her to stay at school instead of being sent away for not paying school fees.

One school kept rabbits not for agricultural purposes or for the school feeding programme but for biology practicals. The rabbits were used in lessons on the various body systems (respiratory system, excretory system, etc.) and were thus dissected to illustrate how the various systems work in the body. Yet another school used the animals they had (cows and chickens) as teaching aids for the few blind children in their student population. The students were taken to touch and feel the animals to get an idea of how they 'looked'.

---

school's own farm was overlooked either because there was very little harvested in 2000 due to the drought or the crop had not yet been harvested due to late planting.

The respondents at the fifty farming schools were also asked about future plans concerning their farming activities. Although about a quarter had no plans at all, the majority did. Fourteen schools (28%) wanted to go into livestock keeping, ten intended to increase the land they had under cultivation, while another five schools wanted to start an irrigation project. The remaining schools simply indicated that they intended to continue with crop cultivation and livestock keeping.

In summary, this chapter has shown that farming in town is not only common among individual households but at schools as well. This applies especially to secondary schools, as they undertook farming almost without exception. Crop cultivation appeared to be the dominant activity; very few schools kept livestock. The labour was usually provided by the students (the agricultural class). The use of inputs was very common and the most conspicuous finding in this respect was that chemicals were applied by all secondary schools and by over half of the primary schools. The produce was in most cases used for the school's feeding programme (lunch), although in some schools the harvest was either sold or taken home by the pupils.

Photo 13 Back-to-back constructed houses in Rhonda Weavers leaving no space for farming.
(Dick Foeken, 1999)

Photo 14 Crop cultivation in a more spacious part of Kwa Rhonda.
(Sam Owuor, 2000)

10

The poor

Introduction

The growth of urban agriculture in sub-Saharan Africa is generally seen as a response to urban poverty. Due to the prolonged economic recession, it has become increasingly difficult for many urban dwellers to find (or indeed keep) a steady job. Meanwhile, people are also faced with steadily decreasing purchasing power. For many urbanites, poor and non-poor alike, the answer lies in diversifying one's livelihood sources. For the large majority, this is only possible in the informal sector. One of these additional livelihood sources is urban farming, which, in a town like Nakuru, is almost exclusively an informal-sector activity. This is not to say that people started farming *only* because of problems with their livelihoods. Urban farming did already exist because "farming is part of our life", as many urban farmers say. But from a primarily cultural phenomenon, urban farming has developed into an economic necessity without which many urban households could not maintain the standard of living they have been used to, or not even survive. The latter applies especially to the urban poor. It is the same poor, however, who may lack the resources to farm, be it access to land or funds for necessary investments. And if they do farm, they may do so on smaller plots, use fewer inputs, receive less assistance and realize less produce than better-off households.

This chapter attempts to analyse how far poverty, on the one hand, and urban farming, on the other, are related in Nakuru town. This has been done by comparing two income categories on a number of issues, as discussed in the previous chapters. A practical definition of 'poor' is used here, namely households with a monthly cash income of Ksh. 5,000 or less.[1] However, since no detailed calculation of household income has been made during the study, it is obvious that this can be no more than a rough indication of 'poverty'. The same applies to the group of households with a monthly cash income of at least Ksh. 10,000, i.e. the medium-to-high income category that is used for purposes of comparison. In the sections that follow, the two income categories are denoted as the *poor* (310 households) and the *non-poor* (106 households), respectively. The basic data are presented in Annex 10, while below it is mainly the differences between the two groups that are highlighted.

Farming

In Chapter 3 (Table 3.3, p. 41) it was seen that the poor were under-represented among urban farmers in Nakuru. This is confirmed by the figures in Table 10.1: among the non-poor, farming in town appeared to be almost three times as common as among the poor. Moreover, this applied to both crop cultivation and livestock keeping. This may to some extent be related to the fact that households were generally larger among the non-poor.[2] It also indicates that the poor faced more serious obstacles to starting farming than the non-poor. Surprisingly, it was not the lack of access to land as such that can explain the difference between the two income groups regarding not-farming because this

Table 10.1 Farming by income class

	poor		non-poor	
	abs.	%	abs.	%
- engaged in urban farming	68	21.9	63	59.4
- engaged in urban crop cultivation	43	13.9	56	52.8
- engaged in urban livestock keeping	43	13.9	37	34.9
- engaged in rural farming	172	55.5	72	67.9
- engaged in rural crop cultivation	172	55.5	71	67.0
- engaged in rural livestock keeping	97	31.3	47	44.3

Source: 1999 survey.

1 With an exchange rate of Ksh. 70 for US$ 1 at the end of 1999, this equals almost US$ 70 per month. With an estimated household size of about 3.5 for the whole study population, this works out at about US$ 20 per person per month, i.e. less than the often used definition of poverty of US$ 1 per person per day. However, this is cash income, excluding 'indirect income sources', such as for instance the monetary value of self-produced food items.

2 See Annex 10, Table A10.1.

obstacle was equally often mentioned by the poor as by the non-poor and regarding both crop cultivation and livestock keeping.[3] Instead, it was to some extent the *combination* of lack of access to land and lack of capital that seemed to explain the poor's under-representation among urban farmers in Nakuru town.

At first sight (Table 10.1), it seems as if the poor were able to compensate for their lack of access to urban land with farming in the rural area, as over half of them indicated having access to rural land. However, a closer look[4] reveals that the rural plots of the poor were on average smaller (3.0 acres) than those of the non-poor (6.2 acres). About half of the poor did *not* use their rural plot themselves, compared to about 30% of the non-poor. Nevertheless, for the large majority of the poor, their rural plot did constitute a food and/or income source and in this they did not differ from the non-poor.

As shown in Chapter 3 (p. 39), there was a substantial decline in the numbers of urban farmers between 1998 and 2000. This tendency appeared to be stronger in the low-income neighbourhoods than in the medium-to-high-income ones. In the former neighbourhoods, the number of crop cultivators declined by almost 40% and the number of livestock keepers by a third. In the medium-to-high-income neighbourhoods, these figures were 15% and 20% respectively. As indicated in Chapter 3, the most likely explanation for this decline was the drought in 1999 and 2000. This is illustrated by responses during the in-depth survey of 2000 to the question of how people had coped with the drought of 1999. From the answers it becomes clear that the availability of water for both crops and animals played a crucial role. Those who practised crop cultivation without irrigation had no, or hardly any, harvest. As one respondent remarked:

> Mostly when there are droughts, the crops are affected and the harvest is small, like the end of last year [1999] and the beginning of this year [2000]. All the kales dried up this year and now we have to buy the food from the market. This has made it difficult for us.

Households that usually irrigated their crops were also confronted with an acute water shortage due to the fact that "the taps dried up", an indication of the widespread water problems Nakuru town faced:

> Because of the lack of constant water supply, I stopped farming, because I was not able to buy water for crop irrigation and household use at the same time.

After the failure of the long rains, some crop cultivators tried to plant again, something that is usually not done in 'normal' years because the second [or short] rains are very unreliable:

---

3 See Annex 10, Table A10.2.

4 See Annex 10, Table A10.3.

> I planted twice during the year because the crop that I planted at the start of the long rains dried so I planted again at the start of the short rains. From this second planting I was able to harvest about one bag of maize but the beans dried once again.

Another respondent also mentioned having planted twice, but the second planting failed as well. Some who did manage a harvest without irrigation were forced to harvest the crop very early to prevent it from completely drying up:

> To avoid losing the whole crop due to the drought we harvested the maize while it was still green.

Still another household harvested both maize and beans while they were still green, which forced them to sell some of it at a low price because green maize has a much lower nutritional value than mature maize and is only fit for roasting.

The drought had an impact on livestock keeping as well. Some respondents blamed the lack of water, and hence the lack of feed, for the loss of some of their animals:

> The drought had very bad effects. Last year's [1999] drought cost me two heifers because of the lack of soft green grass. Moreover, the livestock became so emaciated that they took a longer period to recover and by the time they were recovering, this year's [2000] drought started.

Faced with a lack of easily available grass, many cattle keepers had to buy feed at very high prices because of the general shortage of good feed. At the same time, the animals produced less milk because of their poor condition, so incomes decreased as well:

> Drought made some of my animals to die. Milk production was low and this meant that I sold fewer litres than usual and the animals have not recovered from the effects of the droughts.

Not only large animals like cattle were affected but also small animals like chickens, as a few chicken raisers mentioned:

> When there are droughts, the chickens die from unknown diseases and also from lack of feed. The effects are that we are not able to obtain enough eggs and that we have a reduced stock number.
>
> Because of the lack of water, my chickens got infected. There was nothing I could do about it but to dispose of the whole flock.

Some tried to obtain water and feed from far away, but not always with the envisaged effect:

> The major difficulty was water and lack of feeds. This forced us to go and draw water and also to get feeds from as far as 30 km away. Yet, to cope with the droughts, we had to sell most of the livestock.

Droughts frequently occur in Nakuru and this makes farming in town a risky business, especially for the poor who usually have no constant water supply at their disposal. As a result, many of the poor do not invest very much in their farming activities. This dilemma was summarized by a respondent from Ziwani who lacked water near his plot and who used no fertilizer besides the crop residues left on the land:[5]

> You can start buying things to use on the *shamba* but you know whatever is a key thing? Why use more money if the main thing, water, is lacking? We fully depend on rain and then, if there are no rains, there is a low or no harvest. In such a case when you have used fertilizers or whatever, you make a loss!

Crop cultivation

Table 10.2 shows some characteristics of the plots used for crop cultivation by the two income groups. Although the very small plots (i.e. less than 50 m^2) were somewhat more common among the poor, on the whole, the plot size distribution hardly differed between the two groups. This is related to the fact that over half of the poor crop-cultivating households had a plot *outside* their own compound, where there is more space than *in* the compound (if there is a compound at all). Quite a number of them grew their crops along a road, along the railroad or under a power line. Among the non-poor, the majority had their *shamba* within the compound. Hence, the large majority of the non-poor had the advantage of the plot being nearby. Of the poor, 16% had to walk at least half an hour to reach their *shamba*. Related to the location of the plot is its ownership. Only 20% of the poor said they owned the land, half farmed their landlord's land (which often means the compound around the house they rent), while almost a third grew their crops on land belonging to someone else – the government,

Table 10.2 Summary of urban plot characteristics by income class (%; plots)

		poor (N=46)	non-poor (N=61)
plot size (m^2)	<50	38.0	26.2
	1000+	30.0	31.1
location	in own compound	45.7	68.9
distance to plot	<10 minutes walking	58.0	83.6
ownership of plot	own land	20.4	47.5
	landlord	49.0	41.0
	someone else	30.6	11.5

Source: Annex 10, Table A10.4.

5 Versleijen 2002, p. 52.

a friend or even an unknown landowner. For the large majority of the non-poor, it was either they themselves or their landlord who owned the land.

On average, the poor cultivated a slightly smaller variety of crops than the non-poor: 4.1 and 4.6 crops, respectively. In both groups, maize, beans and kale were by far the most popular crops, the former two especially among the poor and the latter among the non-poor.[6] This suggests that the poor were more inclined to grow basic staple foods than the non-poor. As for the other crops, there were no clear differences between the two groups, although tomatoes and bananas were somewhat more common among the non-poor.

The large majority in the poor as well as the non-poor crop-cultivating households used material inputs.[7] In general, poor and non-poor households showed very little difference regarding their use of certain types of inputs, with the exception of irrigation. The number of households irrigating their crops was more than twice as high among the non-poor as the poor. Tap water is the most common source of irrigation water in Nakuru and, as mentioned above, most poor households lack this provision. Manure and crop residues serving as fertilizers were also somewhat more frequently used by the non-poor. This may be related to the fact that livestock keeping – and cattle in particular – was more common among the non-poor (supply of manure) and that they also had a somewhat larger average plot size (crop residues). Rather surprisingly, chemical inputs were not used any more often by the non-poor, although it is not clear whether they used larger quantities. Finally, the use of improved seedlings was slightly higher among the non-poor.

Yields did not differ between the poor and the non-poor. In absolute figures, the mean harvest was about 350 kg in both groups. Land productivity was also equal, namely 0.37 kg/m^2, which was about the same as the overall average for the whole crop-cultivating population. These findings are in line with findings concerning the use of material inputs. Only irrigation was practised more often among the non-poor, but since 1998 was a fairly ‘normal’ year in terms of rainfall, this played a minor role. The situation was different in the following two years, as seen earlier in this chapter. If measured in that year, average crop yields of the poor would undoubtedly have been much lower than those of the non-poor.

Although there were no substantial differences between the income categories concerning problems with crop cultivation, two things can be mentioned.[8] The non-poor suffered somewhat less from theft than the poor (34% vs. 44%). This can likely be related to the fact that the non-poor were more often able to

---

[6] See Annex 10, Table A10.5.

[7] See Annex 10, Table A10.5.

[8] See Annex 10, Table A10.6.

grow crops within their own compounds. On the other hand, the poor complained a bit less of lack of water for irrigation (5% vs. 18%), despite only a few of them using it. Again, the following years would have seen different answers as far as water availability was concerned.

Livestock keeping
Chapter 5 showed that there is a relationship between household income level, on the one hand, and the types of livestock kept, on the other. This is confirmed by the figures in Table 10.3: only a quarter of the poor kept large livestock, against 70% of the non-poor. Very few of the poor could afford to keep cattle, the most expensive but at the same time the most financially rewarding type of animal. Hence, for most of the poor, small livestock were the only option, usually chickens, while most of the non-poor livestock keepers kept both large and small animals.

There were no clear differences between the poor and the non-poor as far as rearing system was concerned, although the latter kept their animals more often within their own compounds than the former (Table 10.3). Many of the poor had no compound or only a very small one, so free range was the dominant rearing type by necessity. Many of the non-poor, however, and the rich in particular, did have a compound of some size but there was also space outside where the animals could roam around freely.

As for the person responsible, in poor households this was equally divided among the household head and his wife (Table 10.3). In non-poor households, however, it was mainly the spouse, undoubtedly due to the household head

Table 10.3 Summary of livestock-keeping characteristics by income class (%)

		poor (N=43)	non-poor (N=37)
% households keeping	large livestock	23.3	70.3
	small livestock	100.0	86.5
rearing system	within compound	35.9	48.3
	free range	64.1	44.8
person responsible	household head	48.8	27.0
	spouse	48.8	67.0
used hired labour?	yes	14.0	43.2
material inputs	improved breeds/AI	9.3	40.5
	veterinary drugs	23.3	81.1
	feed supplements	39.5	86.5
disposal of animal waste	for own crop cultivation	25.6	70.3
	dumped in the street	48.8	16.2

Source: Annex 10, Table A10.7.

having a job somewhere else. As a result, hiring labour was more common among the non-poor, partly because of a lack of time for the household head in non-poor households and partly because of a lack of money in poor households. The latter is also shown by the use of material inputs. 'Modern' inputs like improved breeds, veterinary drugs and feed supplements were commonly used by the non-poor, while few of the poor could apparently afford them.

There appeared to be a clear difference between the two groups regarding the disposal of their animals' waste (Table 10.3). While a majority of the non-poor used all or part of it for their own crop cultivation, this was done by only a quarter of the poor. At the same time, about half of the poor dumped some or all of their animals' waste in the street, i.e. three times as many as in the non-poor group. These figures reflect to some extent the types of animals kept (non-poor: large livestock that produced large quantities of valuable dung) and the problem of space (poor: no or only small compounds).

Poor and non-poor households showed hardly any difference as far as the various types of problems with livestock keeping are concerned.[9] In both groups, diseases were mentioned by far the most often, particularly in the non-poor group. In general, all problems but one were mentioned more frequently by the non-poor, which could indicate a higher degree of awareness of potential problems in this group, possibly due to the higher value of their animals. The exception is 'lack of space', mentioned by several of the poor but by none of the non-poor, which is not surprising.

Benefits

Food security

The overwhelming majority of both the poor and the non-poor urban farmers said they farmed in town for the food.[10] As for urban crop cultivation, this was considered to be a major food source for a fifth of the poor, which was a slightly higher percentage than for the non-poor. For the latter, cultivating crops in town was more of an additional food source. For some, crop cultivation in town was so important that they "could not survive without it". Not surprisingly, this was more common among the poor than among the non-poor. The importance of urban livestock keeping as a food source was equal in both income groups: for about half, the practice was an additional food source, while 16% of the respondents mentioned it to be a major food source.

---

9 See Annex 10, Table A10.8.

10 See Annex 10, Table A10.9.

During 1998, there were no serious food problems in the two groups, because the large majority had "always had enough to eat" (Table 10.4). However, those who did face food problems were mainly found among the poor. Although generally speaking, purchased food was the most important food source for the majority among both the poor and non-poor urban farmers, for many, one's own urban production was also among the "most important food sources", while quite a number relied on their rural produce as well. In general, own food production, be it urban or rural, was more important for the non-poor. However, the importance of urban crop cultivation for the poor is shown by the fact that for a majority (60%) their own urban food production constituted at least half of their household's food consumption – a figure twice as high as that among non-poor urban crop cultivators.

Table 10.4 Urban farmers: summary of general food security issues by year and income class (%)

		1998		1999	
		poor	non-poor	poor	non-poor
		(N=68)	*(N=63)*	*(N=13)*	*N=31)*
always enough to eat?	yes, always	88.2	96.8	23.1	75.9
				(N=13)	*(N=27)*
*most important food source(s)**	urban production	41.2	50.8	15.4	35.5
	purchased	85.3	74.6	100.0	96.8
	rural production	25.0	35.0	-.-	16.1
contribution of urban crop cultivation		*(N=43)*	*(N=56)*	*(N=10)*	*(N=26)*
to household	at least half of the food	60.4	32.2	20.0	23.1
food consumption	small proportion or less	25.6	43.6	60.0	53.8

* Total exceeds 100%.
Source: Annex 10, Table A10.10.

The two right-hand columns of Table 10.4 give an indication of the serious effect of drought on the food security situation of the households in Nakuru town, for the poor in particular. Even though (1) the figures can only be seen as rough approximations (in qualitative terms) of the households' food security, and (2) the 1999 numbers of the poor urban farmers are small, the 1999 figures show notable differences with those for 1998. The percentage of poor households indicating that they had "always enough to eat" had dropped from almost 90% in 1998 to only 23% in 1999. The figures show that this was to a large extent caused by the failure of the urban as well as rural harvest, as the percentages of poor households for whom these were important food sources, declined equally. And although all these percentages in 1999 were lower among the non-poor households as well, these drops were far less dramatic than among the poor. As a result, dependence on purchased food increased considerably. However, the figures also indicate that even though 100% of poor households relied on purchased food, they could not afford to purchase all they needed, witness the fact that only a quarter of them always had enough to eat. In other words, the remark that they "could not survive" without urban (and rural) farming, would appear to be no exaggeration.

Income

From a commercial point of view, livestock keeping was more important than crop cultivation, and this applied to both income groups.[11] Moreover, the percentage of respondents indicating whether urban livestock keeping was an income source – either a major or an additional source – was about the same in the two groups. With urban crop cultivation, the situation was somewhat different, however. Growing crops in town not only for food but also for income was more common among the poor than among the non-poor.[12] For many of the poor, livestock keeping to obtain an income is beyond their reach because of the investment costs involved, so they try to earn extra income from crop cultivation.

As mentioned in Chapter 6, urban farming not only provides the household with a direct income from sales but also with an indirect income by saving on the cost for buying food. How important this indirect income – and hence the urban plot – is for a poor household can be illustrated by the case of Baba and Mama Christopher.[13] They cultivated a plot of 50 m^2 along the railway line and belonging to Kenya Railways, but suddenly the "railway boss" forbade them to use the plot any longer because of a fuel tank nearby. The plot was very important to them because:

> You know if you manage to grow your own food for several months per year, then you can educate your children from your salary.

Things worsened because around the same time, both Baba and Mama Christopher became ill, putting an extra burden on the household, as Baba Christopher explained:

---

11 See Annex 10, Table A10.9.

12 This is shown in Table A10.9 by the percentages of respondents indicating 'needed income' and/or 'to diversify income' under 'reasons to farm in town', as well as the percentages 'could not survive without it', 'major income source' and 'additional income source' under "importance of urban farming".

13 See Versleijen 2002: 36-37.

> My children will have to drop out of school since I cannot pay the school fees.[14] You know, right now all the money goes to food and medical bills (…). So what will we do? I cannot educate my children anymore. At least when I have a small *shamba*, I can get most of the food from there and then I can put my children back in school again.

To get at least some 'free' food, Mama Christopher used to collect 'wild' vegetables (*mchicha*, *managu*, *saget*, etc.) from a small field. This field was used by a few other women as well who, like Baba and Mama Christopher, lacked a *shamba* on their own. According to Versleijen (2002: 37), the "gathering of 'wild plants'[15] is a practice done by more women in low-income households". Mama Christopher also used other spots to collect them:

> You know, things become difficult now. Each day I'm looking for vegetables to eat (…). I look for vegetables everywhere these days. You know, the place I used to pick them, they are not there anymore, because it was only a small place and I went there frequently. So now I just go looking at the side of the road or wherever I think I might find them. But you know, it's a lot of work, I'm spending a lot of time on it. But what else can we do? There is nothing else we can do than pray to God.

Support

As indicated in Chapter 7, very few of the crop cultivators in Nakuru town received technical support, but quite a number of the livestock keepers did. This was mainly the non-poor (Table 10.5). Very few of the poor received assistance and only two of these received assistance from a professional. The difference between the two groups is indirectly related to the differences in welfare level. The poor can generally not afford to buy large livestock and it was mainly for these animals – cattle in particular – that assistance was provided or called for.

Table 10.5 Assistance with livestock keeping by income class (%)

	poor (N=43)	non-poor (N=37)
households receiving assistance with livestock keeping	14.0	54.1
idem, from professionals	4.7	43.2

Source: 1999 survey.

---

14 Actually, the school-aged children were sent home several times and two of them were told not to return for the rest of the term.

15 Wild plants include various types of *mchicha* and other greens that can be found along roadsides or in small open areas.

A positive development in this respect is the credit programme run by ECLOF Kenya, because it specifically focuses on the poor and the marginalized. As described in Chapter 7, some people did benefit substantially from participation in the programme. However, according to Theresa Moyo,[16] micro-finance systems like this are under threat due to HIV/Aids. Sick people cannot work and the borrowed money is quickly spent on medicines or a funeral. The result is that loans cannot be paid off.

Environment

Generally speaking, the poor live in more seriously polluted parts of Nakuru town than the non-poor. This is illustrated by the numerous garbage dumps that can be observed in the streets of the low-income neighbourhoods. It is also indicated by the heavy metal concentrations as presented in Tables A8.1 and A8.2 (Annex 8). Among the twelve sampling sites, there was one in a high-income area (Milimani) and there were three in medium-income areas (Lanet Free Area, Lanet and Kabachia). All other sites were located in low-income areas, with the exception of the dump, although the ones who were farming there came from low-income households in the immediate vicinity. Overall, the soils as well as the plants in the low-income sites and at the dump appeared to contain higher concentrations of heavy metals than in the sites in the medium-income and high-income areas. To some extent, Kabachia is an exception, but the comparatively high concentrations of zinc in the soil and lead in the plants were due to the burning of some tires close to the selected site.

Schools

Although in many ways the poor are in a disadvantaged position, in theory, they might benefit from school farming and/or school meals. School farming provides produce that can either be used for feeding programmes at school or that can be taken home by the children involved, thus providing additional food for their households. In schools where a decent lunch is provided, children can benefit as well, especially those from poor households. Hence, in terms of 'poverty alleviation', school farming and school feeding are particularly important in low-income neighbourhoods. Although the school survey done by Odera did not focus on that particular aspect, it is nevertheless possible to get at

---

[16] SAMCAF, a network of micro-finance organizations in Southern Africa. See interview with Ms Moyo in *Trouw*, 17 April 2004.

least some idea as to how far school farming and feeding are related to the income level of the neighbourhood in which the schools are located. Table 10.6 provides some figures.

Table 10.6 School farming and feeding by income level of neighbourhood (%)

		'lower'* (N=23)	'higher'* (N=21)
% of schools engaged in farming		47.8	66.7
plot size**	smaller than 0.5 acre	63.6	8.3
	bigger than 1.0 acre	18.2	75.0
% of schools providing lunch		43.5	66.7

* 'Lower' consists of "area income level" categories 'low' and 'low to medium' in Table A9.1 'Higher' consists of "area income level" categories 'medium to high' and 'high' in Table A9.1. The classification is made with the aid of MCN (1999) and the map of Nakuru Municipality (Nairobi: Survey of Kenya, 1998).

** Only schools performing crop cultivation: 11 in the 'lower income' and 12 in the 'higher income' category.

Source: 1999 survey.

School farming appeared to be somewhat less common in lower-income than in higher-income neighbourhoods. Perhaps more importantly, the plots used for crop cultivation by schools in lower-income neighbourhoods were on average (much) smaller than those in higher-income neighbourhoods. That is, however, not simply because school compounds are always (much) smaller in lower-income neighbourhoods. For instance, three of the eleven crop cultivating schools in the lower-income neighbourhoods had a 'surplus' of more than five acres of land within their compound (including the school buildings), indicating the potential to expand their farming activities.

Finally, the picture regarding school feeding appears to be the same as with school farming (Table 10.6). In schools located in the lower-income neighbourhoods, the provision of lunch for the children was somewhat less common than in schools in the higher-income neighbourhoods. Based on these crude data, one can conclude that in neighbourhoods where school farming and school feeding were most needed, it was less practiced.

Gender

So far, this chapter has focused entirely on the poor in relation to urban farming. However, in the livelihoods approach, another important social variable is considered to be of particular importance: gender.[17] Although the study did not focus especially on the gender aspects of urban agriculture, from the collected data it is possible to present some findings on differences between male- and female-headed households. Moreover, some findings related to the 'person responsible for crop cultivation' – i.e. the male head, the wife of the male head and the female head – were already presented in Chapter 4.

The reason to include the section on gender in the present chapter is largely because 94% of the female-headed households belonged to the "low-income" and "very-low-income" categories (25% and 69%, respectively). Even so, 84% of the female-headed households practising urban farming belonged to these low-income groups. The comparative percentages for the male-headed households were 79% and 67% respectively. In other words, both among the whole study population and among the urban farmers, the female-headed households were on average somewhat poorer than the male-headed households.

Twenty-seven per cent of the female-headed households could be classified as urban farmers, which was a lower percentage than for male-headed households (37%). The same trend applied to the two components of urban farming: urban crop cultivation (20% and 29%, respectively) and urban livestock keeping (13% and 22%, respectively). Put differently, female-headed households were somewhat underrepresented among urban farmers.

The 23 female-headed households cultivating crops in Nakuru town did so on a plot averaging 688 m^2, which was considerably smaller than the average plot of their male-headed counterparts (1026 m^2). Almost half of the plots belonging to female-headed households were smaller than 100 m^2.[18] Almost 90% of the plots were located in their own compounds or within the estate where the farmers were living, which was on the whole a slightly higher percentage than the plots of the male-headed households (75%). Hence, 85% of the plots were owned by either the female head herself or a landlord. Only one female head (4%) cultivated crops on a piece of government land. Among the male heads, farming on government land was somewhat more common (16%).

As shown in Chapter 4, compared to women, men were more inclined to cultivate traditional staple crops such as maize and beans. For vegetables, it was the other way around. In general, women cultivated a larger variety of crops. Perhaps the most conspicuous differences between men and women concerned the use of inputs for crop cultivation (as was shown in Annex 4, Table A4.3). Men not only used more inputs, but also more 'modern' inputs (chemicals and improved seeds). While over half of female heads used no modern inputs at all, this applied to less than a quarter of the male heads, while of the latter, almost

---

[17] See the section "Theoretical considerations" in Chapter 1.

[18] See Annex 10, Table A10.11.

half used at least two modern inputs, against only one female head. For chemicals alone, the differences are even more striking, as 90% of the female heads used no chemicals at all. Irrigation, on the other hand, was much more common among the female heads and also among the wives of the male heads. As for hiring additional labour, few female heads (17%) did so, at least fewer than the in the male-headed households (29%). All this helps to explain the very low total yields in the female-headed households in comparison with the male-headed households, not only in absolute terms (almost 140 kg in the female-headed households and almost 300 kg in the male-headed households) but also when measured in terms of land productivity (0.20 kg/m^2 and 0.30 kg/m^2, respectively). The latter is the more remarkable as the average plot size of the female household heads was much smaller than that of male heads.[19]

In summary, this chapter shows that the poor are in various ways disadvantaged when it comes to urban farming. Compared to the non-poor, they were quite under-represented among urban farmers, few of them used irrigation for their crops, very few kept large livestock and equally few received technical assistance. Female-headed households – a special category among the poor – were even worse off because they had smaller plots, used fewer inputs and had much smaller harvests than male-headed households. In terms of food security, the poor are much more vulnerable than the non-poor, which was dramatically shown in 1999 when there was a very bad harvest due to a lack of rainfall. An additional disadvantage is that the poor tend to live in the more polluted areas of town. Even school farming and school feeding – which is potentially beneficial for children from poor households – appeared to be less common in low-income neighbourhoods compared to higher-income areas.

[19] See Table 4.5 (p. 59) for the relationship between plot size and harvest per m^2.

Photo 15 Dairy cow in zero-grazing in Rhonda Weavers.
(Sam Owuor, 2000)

Photo 16 Rabbit keeping in Lanet.
(Sam Owuor, 2000)

11

Summary and conclusions

Introduction

Samuel and Pauline came from Bomet to Nakuru in 1987 and settled in Rhonda Kaptembwa, where they still live today.[1] *They have always cultivated crops and kept cows in their compound on a plot of about 60x30 metres. In 1999 they grew maize and beans solely for self-consumption and Napier grass for their cows. They weeded twice and used chemical pesticides. Due to the drought and because they did not irrigate their crops, their harvest was modest: some maize cobs were picked raw and roasted, while only a few kilograms of beans were harvested. The dried maize stalks were fed to the cows and Napier grass was cut whenever it reached a certain height. They did not sell anything but gave away some of their crops to friends. Pauline is responsible for the crops and at peak times she works on the* shamba *all day, with some assistance from a nephew. One day she also hired a local person to weed, which cost her Ksh. 300. The crop residues were used as fodder for the cows and the animal dung as manure for the crops. Farming activities are important for her because "it subsidises". She would not stop cultivating even if the household had sufficient income to ensure a decent standard of living. Pauline is convinced that she could produce more crops if she was able to irrigate the crops. The local government could assist by providing a water supply.*

---

1 Rhonda Kaptembwa is a low-income, high-density residential area in the southwest of Nakuru town (see Map A1.1 on p. 176).

This brief description of the urban farming activities of Samuel and Pauline includes many of the aspects dealt with in this book: food security (produce for self-consumption), income ("it subsidises"), employment (family and hired labour), environment (chemical inputs, animal waste for crop cultivation), constraints (rainfall) and the role of the local authorities (provision of water supply). In the next section, the main findings related to these aspects are summarized for the whole study population as well as for the sub-studies. The following section discusses some of the main findings in relation to the theoretical considerations outlined in the first chapter. The final section deals with the future of urban farming in Nakuru town in the context of recent policy developments in the municipality.

Summary of findings

Samuel and Pauline's household was one of an estimated 25,000 households in Nakuru town that practised urban farming in the late 1990s. This figure accounts for about a third of Nakuru's households. These 25,000 were not equally distributed over the town's population. Compared with non-urban-farming households, households performing urban farming were generally larger, i.e. with more mouths to feed. Another difference between farmers and non-farmers was household welfare level (at least measured in terms of monthly cash income): the poor were under-represented among urban-farming households. Yet, low-income households were the largest group among urban farmers and urban agriculture is very important as a food and income source for this group. The dominant reason for the non-farmers not to farm in town was lack of access to urban land, followed by other considerations among which lack of capital was the most important. For some, there was no need to undertake urban farming because of access to rural land, although for many others this was not a reason not to farm in town as well.

Nakuru households cultivate crops whenever they have access to a suitable piece of land. For the majority, this land lies within their own compound. However, no less than 40% of the plots were located elsewhere, mostly being government land or 'undeveloped' private land. The average plot was almost 1,000 m^2 (0.25 acres), but plots varied greatly in size. The total area under crops in the built-up area of Nakuru amounted to more than 5,000 acres (2,000 ha) in 1998. Most plots had been put into use after 1990, many even after 1995, indicating that crop cultivation *at this scale* is a fairly recent phenomenon in Nakuru.

A whole range of crops was cultivated by the Nakuru households, but these were overwhelmingly food crops for self-consumption. Mixed cropping, with traditional, simple equipment and carried out with family labour, was very common. Almost all crop cultivators applied material inputs and the use of organic fertilizer was common. About half of the cultivators used chemical inputs. The same applied to irrigation, mostly tap water. Yields were quite low, partly depending on the size of the plot (the smaller the plot the higher the land productivity) and the number of inputs, irrigation in particular. Very few crop cultivators received any technical assistance. Apart from typical problems related to farming, crop theft was a major urban-related constraint for cultivators. Despite this, in 1998 (a 'normal' year in terms of rainfall) some six million kg of crops were harvested, which contributed about 22% to the energy requirements of the producing households and about 8% of the requirements of the whole population of Nakuru.

Although the absolute number of households keeping large livestock in town was relatively small, the total number of cattle, sheep, goats and pigs in the built-up areas of Nakuru town could be estimated at some 25,000 in 1998. Small livestock (mainly chicken) were more common, with numbers totalling some 380,000. Large livestock were kept for both self-consumption and commercial purposes, small livestock more for self-consumption. There appeared to be some relationship between type of livestock and income class, as large livestock were more commonly found among higher-income households.

Livestock were partly kept in people's own compounds and partly free range. Small livestock roamed freely somewhat more often than large livestock. In general, large livestock received more attention than small livestock, at least when one considers the percentage of households using certain inputs. In contrast with crop cultivation, quite a number of livestock keepers had received technical assistance, especially those keeping cattle. However, the death rate among animals was equally high among both those households that had received assistance and those that had not. Related to this, 'disease' was by far the most frequently mentioned constraint.

An important environmental issue in town is related to the waste from livestock. About a third of livestock keepers dumped all or part of their animals' waste in the street, but this was more common for small-livestock waste. Most of the waste, however, was used for crop cultivation, either by the livestock keepers themselves or by their neighbours.

For most households practising urban farming, the need for (additional) food was the main reason to engage in it. This applied more to crop cultivation than to livestock keeping. Livestock were also kept to obtain additional income and/or to diversify income sources. As perceived by the respondents, about 40% of crop cultivators stated that their urban cultivation constituted at least half of the food they consumed (in 1998). And the large majority of those engaged in urban farming said that it formed an *additional* food and/or income source, while for about a quarter it was the *major* source. The level of household food security was somewhat better in farming households than in non-farming households. Probably partly as a result of this, the average growth in height of farmers' children was somewhat better than that of non-farmers' children. The year 1999 was quite different, however. As a result of the drought, there was little harvest so urban farmers had to purchase much more food than in 'normal' years, thus saving less money for other expenditures.

Another benefit of urban farming – and of particular importance in times of retrenchments and increasing unemployment – is the creation of work. The fact that 25,000 households were engaged in urban farming implies that at least the same number of persons were in some way involved in farming. Moreover, for about a fifth of them it was a full-time job. In addition, about 8,500 persons found work as labourers, either casually (in crop cultivation) or more or less permanently (in livestock keeping).

Support for urban farmers was not widespread and crop cultivators were hardly ever visited by professional officers. There was more support for livestock keepers, especially those with cattle. The study showed that assistance from professional officers positively influenced rearing systems (more zero-grazing), inputs (more modern inputs) and the disposal of animal waste (more used for own crop cultivation). Some farmers in Nakuru participated in the Agriculture and Rural Development Programme (ARDP). This sub-study compared participants and non-participants and showed that the programme had a positive impact on sales of cattle and milk and thus on the income situation of the households involved. In 2001, the Ecumenical Church Loan Fund (ECLOF Kenya) started offering small loans in Nakuru. Among the recipients were some urban farmers. The four (randomly chosen) case studies presented in this book were all success stories.

Urban farming is always related to the urban environment, mostly in a negative way because the activity is supposed to cause all kinds of damage and pollution. Although livestock keeping is usually considered to be worse for the urban environment than crop cultivation, another sub-study showed that the awareness among Nakuru crop cultivators regarding the possible damage of their activities for the urban environment was greater than among livestock keepers. A large majority of the livestock keepers were satisfied with the way they disposed of their animals' waste but quite a number admitted that their neighbours did complain.

Among non-farmers as well as officials, perceptions regarding the relationships between urban agriculture and the urban environment in Nakuru were on the whole rather negative. Yet, both groups were generally convinced that urban farming was important for the urban food supply and would be more acceptable if it were better controlled. Particularly the non-farmers saw a role for the government in tackling the (perceived) negative environmental impact of urban farming.

Measurements of heavy-metal concentrations in soils, water and plants at twelve selected sites showed that concentrations were relatively high at sites where sewage water was used for irrigation and at the dump (where a thin layer of soil covers the garbage and where crops are cultivated on a large scale). Crops growing near busy roads also contained larger concentrations.

Farming in Nakuru Municipality takes place not only in individual households but is also undertaken by institutions. School farming is the most important type of institutional farming in Nakuru and in 2000, about half of the primary and 90% of the secondary schools were engaged in it. Crop cultivation was the dominant activity, on plots of an average size of 1.7 acres. The use of inputs was widespread, but the two school types differed considerably in this respect. Manure was very common at primary schools, but was hardly used by secondary schools. On the other hand, the use of chemicals (including pesticides and insecticides) was widely used at secondary schools, to a much larger degree than at primary schools. In 60% of schools, produce was used for the school meals, particularly at secondary schools.

As mentioned above, although urban farming was found among all income classes in Nakuru town and the poorest households were quite under-represented, the latter still formed the largest group among Nakuru's urban-farming population. Plot sizes did not differ very much between income categories. However, there was an important difference as far as yields from crop cultivation were concerned, as high-income households realized a much bigger harvest than low-income households. This can partly be related to the use of material inputs and irrigation, which were more common among high-income households, and partly to the labour factor, as high-income households were able to hire additional labour when needed. Theft of crops was more of a problem for low-income households, which is likely to be due to the fact that their plots were more often located outside their compound.

Livestock keeping was found among all income categories, although relatively few of the poor households had large livestock (mainly cattle). They also used fewer inputs for their animals, particularly such 'modern' inputs as improved breeds, veterinary drugs and feed supplements. Moreover, they hired less additional labour and received hardly any professional assistance compared to their better-off colleagues. Dealing with animal waste was more of a problem for low-income households because the habit of dumping in the street was more common among them than among higher-income households.

The benefits of urban farming for poor households were dramatically shown in an indirect way. Referring to the 'normal' (rainfall) year of 1998, 60% of these households indicated that their own urban production contributed at least half of the food they needed. The following year was a dry year and this percentage dropped to 20%. And the percentage of poor households indicating that they "always had enough to eat" dropped from almost 90% in 1998 to less than 25% in 1999. In comparison, among the non-poor, the difference between the two years was quite small.

Finally, gender differences in relation to urban farming were found. In general, female-headed households were somewhat under-represented among both urban crop cultivators and urban livestock keepers. The same applied to rural farming. Urban plots were generally smaller among female-headed households. Although these women cultivated a wider variety of crops, their yields were much lower compared to those in the male-headed households, which can be attributed to their very limited use of modern inputs. Keeping large livestock was rare among female-headed households, but otherwise there were no major differences regarding urban livestock keeping between male-headed and female-headed households. The same applied to the various indicators of the benefits of urban farming. However, those data referred to 1998. Since the large majority of the (urban-farming) female-headed households had low incomes, they probably suffered in 1999 from a lack of food due to very small harvests.

Theoretical reflections

The strong growth of urban agriculture in sub-Saharan Africa is related to the economic crisis that has prevailed in most African countries. Also in Nakuru, the increase in crop cultivation and livestock keeping in town can only be assessed in the context of decreasing household purchasing power and increases in the cost of living. As for growing crops, most plots have been put under cultivation since 1990 when the economic recession began to be seriously felt in Kenya, especially in urban areas. The reasons given for turning to this activity confirm that for most people it is a way to secure their food supply and reduce costs on food purchases so that other important expenditures, such as school fees, can (still) be paid.

Thus, the increase in urban farming can be considered as a response to adverse economic circumstances. However, it is by no means the only response. People react in a number of ways, of which the diversification of income sources is undoubtedly the most notable.[2] A wide range of activities, all in the informal sector, are being undertaken, including own food production (urban and/or rural agriculture), manual jobs, petty trade and, especially in the case of

2 See, for example, Bigsten & Kayizzi-Mugerwa 1992; Ellis 2000; de Haan & Zoomers 2003; Kaag *et al.* 2004.

the very poor, prostitution and theft. In addition, social networks are being exploited, with examples being women's groups, merry-go-round groups, ethnically based groups, etc. As for Nakuru's population, Owuor & Foeken (2006) described the livelihood strategies of five low-income households. The number of income-generating activities ranged from four to seven, the mode being six (in three of the five households). All five households practiced urban farming, but it is very clear that urban farming is just one of these households' livelihood strategies.

It would be wrong to consider farming in Nakuru town only as a response to the economic crisis. Quite a number of the Nakuru farmers stressed that farming "is part of their life". In other words, it is also a cultural phenomenon: whenever you have (access to) a piece of land, you do not leave it idle. However, it is the strong *growth* of the phenomenon, including farming on all kinds of plots that do not belong to the people using it, that can only be explained by the economic stress these people are facing. For those with high incomes, it is a way of, for example, still being able to have a car. For the medium-income group, it may be a means to allow the purchase of certain consumer goods. For the low-income group, it is a means to, for instance, still be able to pay the school fees for their children. For the very-low-income group, it can mean survival alone. To summarize, for all households, it is one of the means of maintaining a certain standard of living. Or, as some respondents stated, "it subsidises my income".

Despite the importance of farming in town for the households involved, almost two-thirds of Nakuru households did not undertake any agricultural activities in 1998. For the large majority of these people "no access to urban land" was the main reason. Land, indeed, is the basic resource for urban farming: it has to be available, it has to be accessible and it has to be suitable (Mubvami *et al.* 2003). To start with the latter, suitability, the volcanic soils and relative flatness of the area are favourable circumstances for crop cultivation in Nakuru town. In general, the availability of open spaces has decreased over the past decades, which has caused the cost of renting a piece of land to rise to such an extent that low-income households can no longer afford it. So, for poor households with no compound of their own they can use for growing some crops and/or keep some animals, the accessibility of land has been reduced because of the costs involved. For over a quarter of the non-farmers, "lack of capital" was another important reason not to farm. It is therefore not surprising that the category of very-low-income households was seriously under-represented among the urban farmers in Nakuru.

Still, about a fifth of the very-low-income households did farm in Nakuru town. Compared with non-poor farmers, however, they had less access to all kinds of other resources necessary to optimise their farming activities, including water. In an unreliable climatic region like that of Nakuru, water is a scarce resource, so irrigation is vital to ensure a good annual harvest. Only a quarter of poor farmers had access to irrigation water, compared to almost 60% of the non-poor. As a result, the food security effect of the drought during 1999 was seriously felt by poor farming households. Some people tried to solve the water problem by tapping the sewers, causing soils and crops to be polluted with high concentrations of heavy metals. The use of sewage water for crop irrigation was perceived by the local authorities as the cause of an outbreak of cholera in the town in 2004 and led to the slashing of crops, thus denying these (low-income) households the need for further access to irrigation water.

Keeping livestock in town – and large livestock in particular – requires financial resources for the necessary investments in animals, shelter, drugs, feed supplements, labour, etc. Poor households usually lack these resources, which explains why so few Nakuru households kept large animals. Instead, almost all the poor livestock keepers in Nakuru had chickens. As a result, access to support from professional agricultural officers was also denied to poor livestock keepers because extension services were almost solely directed at helping farmers with cattle. Thus, nearly all the poor households were excluded from making an income from sales of milk and cattle, which can be quite rewarding as the example of Baba Josephine showed (Annex 6, Table A6.5).

In the Kenyan context, access to land is not limited to urban land; on the contrary, access to rural land is much more common. This was also the case in Nakuru, where the percentage of households practising rural farming (by Nakuru townspeople) was almost twice as high as those engaging in urban farming. The importance of rural food and income sources for people in town was the topic of a separate study (Owuor 2006) in the context of the larger Nakuru research project. The study showed (1) that the large majority (85%) of the households in Nakuru benefited from the rural farming activities by either themselves or their rural kin, (2) that in terms of production, rural farming (by urban households) is more important than urban farming, and (3) that the dependency of urban households on rural food and income sources has increased over the past few decades. These findings are in line with recent observations in the livelihood literature that (urban) livelihoods have become increasingly multi-local or multi-spatial, i.e. households have an economic foothold in both urban and rural areas.[3]

Access to urban land is limited for the urban poor because of financial constraints. For rural land, however, access is usually through inheritance. As a result, the percentage of poor households engaged in rural farming was only slightly lower than for the non-poor (some of whom had bought rural plots not

[3] See, for example, de Haan & Zoomers 2003; on Nakuru, see Foeken & Owuor 2001, Owuor 2006, and Owuor & Foeken 2006.

far from Nakuru town). Thus, many poor households in Nakuru that are excluded from farming in town can compensate by obtaining food from the rural home. Yet, it would be wrong to consider urban farming and rural farming as substitutes for each other, as the latter is done because of the opportunity that arises through inheritance, while the former should be seen as a necessity (on 'opportunity' and 'necessity', see, for example, Tellegen 1997).

Another recent observation in the livelihood discourse is that the choice of activities and strategies depends on a number of household and individual characteristics, particularly income and gender (Beall 2002; Kaag *et al.* 2004). The income issue was dealt with above, but the gender aspect is still quite unrecorded in livelihood studies, including urban agriculture. According to Kanji (1995), women in particular increase their informal income-generating activities in order to cope with the declining purchasing power of their household's income. Moreover, these activities are generally concentrated in or near their urban homes (see, for example, Wallman 1996), which was confirmed by Owuor & Foeken (2006) in the context of Nakuru. Farming in town is one such activity. In the literature, urban agriculture in eastern Africa – if not sub-Saharan Africa as a whole – is usually seen as women's business (see, for example, Obudho & Foeken 1999), but it would be an exaggeration to see it solely that way. In about 30% of the farming households in Nakuru, a man was responsible for the activity, there being no difference between crop cultivation and livestock keeping.

Moreover, although the present study's focus was not on gender aspects, some differences between female-headed and male-headed households were noted. Female-headed households were under-represented among urban farmers, had smaller urban plots, used fewer modern inputs for crop cultivation and had therefore much lower yields. These differences are related to the same constraints noted for low-income households (lack of access to land and to capital), which is not surprising as the large majority of female-headed households fell in the low- and very-low-income categories. Yet, there *are* female-headed households that manage to successfully tackle poverty through urban farming, as was shown by the case of Grace Wanjiru who obtained a loan from ECLOF Kenya. By exploiting her social resources – by forming a self-help group as required by ECLOF – and her human resources (skills, experience, labour), she managed to set up a thriving egg business, thus considerably improving her household's income situation.

Policy and the future of urban farming in Nakuru

The importance of farming within the boundaries of Nakuru Municipality has been clearly demonstrated by the studies that were done in the context of the Nakuru Urban Agriculture Research Project (NUAP) presented in this book. The figures concerning the numbers of crop cultivators and livestock keepers as well as the amount of crops produced speak for themselves. Potentially, however, crop yields could be much higher. Plot size proved to be a major determinant of productivity: the smaller the plot, the higher the yield. In other words, if larger plots could be as productive as smaller plots, the municipality could produce a substantial part of the food crops it needs within its own boundaries. However, as shown by this study, more irrigation, inputs, labour and technical assistance are required. In general, this involves more capital, which is a major constraint for low-income households and for female-headed households in particular. More research on the technical aspects of farming in town is necessary to obtain detailed knowledge about present farming practices and to raise productivity.

How productive the sector can be was shown by the example of the Cuban capital of Havana (see Gonzalez Novo & Murphy 2000). Faced with a serious economic crisis, the Cuban government started to actively encourage urban farming, based on the slogan "Production by the neighbourhood for the neighbourhood". An urban agriculture programme was launched integrating access to land, extension services, research and technology development, new supply stores for small farmers, new marketing schemes, and the organisation of selling points for urban producers. The so-called *organopónicos* and *huertos intensivos* (intensive gardens) – raised beds on which horticultural crops are produced in an entirely organic way – have been particularly successful. In 1999, the 773 units covered an area of 386 hectares and produced over 50 million kilograms of crops (i.e. 14 kg/m^2). By comparison, the 2,000 hectares of land used for crop cultivation in the built-up area of Nakuru town only produced some 6 million kilograms of crops in 1998 (a normal year in terms of rainfall). Trials closer to home – in the Tanzanian cities of Dar es Salaam, Arusha and Dodoma – have shown that yields of leafy vegetables can be raised substantially to a level much higher than that produced in Nakuru at present (Jacobi 1997).

Until recently, urban planners have tended to consider urban farming as a temporary feature. According to the *Strategic Nakuru Structure Plan*, this vision prevails in Nakuru as well: "Economically, urban agriculture is a transitory activity which eventually gives way to more traditional urban functions" (MCN 1999: 44). Besides the fact that agriculture has always been part of any urban economy (and in that sense can be seen as a traditional urban function), it should be realized that "although some forms of urban agriculture are based on temporal use of vacant lands, urban agriculture as such is a permanent feature of many cities in developing (…) countries and is thus an important component for sustainable city development" (de Zeeuw *et al.* 2000: 162).

How urban agriculture can be integrated in urban planning policies was the subject of a five-day workshop held in Havana in October 1999 (Bakker *et al.* 2000). The outcome of the workshop was an "outline of a policy framework for urban agriculture", i.e. urban agriculture should be integrated in (1) urban land use policy, (2) urban food security and health policy, and (3) in urban environmental policy (de Zeeuw *et al.* 2000: 164-173).

(1) Integration of urban agriculture in urban land use policy implies the removal of legal restrictions imposed on farming in town and the integration of agriculture in urban development planning (de Zeeuw *et al.* 2000: 165-168). In the context of the *Localising Agenda 21: Action Planning for Sustainable Urban Development* programme, a serious attempt to develop urban planning and management in Nakuru was undertaken in the mid-1990s. Funded by Belgian aid money, a town-planning unit was created and the so-called *Strategic Nakuru Structure Plan* was set up and eventually approved in 2001. Although the Structure Plan was described as "the blueprint for urban sustainable development" (Mwangi 2001: 17), any reference to urban agriculture, except the above-mentioned remark on the transitional character of the activity, was conspicuously lacking. However, there was awareness among key officials that urban agriculture should indeed be part of the Nakuru planning process. This was the reason that NUAP was very much welcomed by these officials.

By the end of the 1990s, farming in town was officially illegal but was tolerated by the authorities. The main legal control mechanism was the Public Health Act: any farming activity considered being detrimental to public health and/or safety or about which people complained was dealt with.[4] This concerned mainly livestock confiscated because it caused a nuisance to neighbours or the wider community (e.g. traffic accidents). Destruction of crops was rare, although it did happen in 2004 (see Box 8.1). While completing this book, new by-laws regarding urban agriculture in Nakuru Municipality were in the making.

Official attitudes regarding urban agriculture may differ from one municipality to the other, from one local department to another and even within departments (Mushamba 2002). This was illustrated by a study in Tanzania, where despite favourable national legislation regarding urban agriculture, the local policy in one town (Morogoro) on this subject was positive but negative in another (Mbeya), although the opinions of officials in the latter town differed in

4 Personal communication, Mr. S.C. Kiarie, Public Health Officer, Municipal Council of Nakuru, 7 September 1998.

this respect (Foeken *et al.* 2004: 124-127). In Nakuru, a comparable situation existed from the end of the 1990s until recently. This emerged during a workshop in November 2002 when the results of the various NUAP studies were presented and discussed by the various stakeholders in Nakuru (see Annex 11). Although the official policy in Nakuru had long been one of toleration, at the beginning of the workshop the then Director of Environment appeared to be firmly opposed to any form of agriculture in town. After two days however, his view of farming in Nakuru town had become more balanced. His colleague of the Department of Housing said the workshop had been "an eye-opener" stressing that "we need to revise our housing policy", i.e. provide new municipal houses with a compound so that the inhabitants can produce some of their own food. A few years later, in 2005, the Nakuru Department of Environment had become an active promoter of urban farming, i.e. as long as the activity is carried out in an environmentally friendly way. The workshop was, at least in part, responsible for this change in attitude and policy.[5]

(2) Integration of urban agriculture in urban food security and health policy implies improved access for urban farmers to agricultural research, technical assistance and credit services, improved systems of input supply and product distribution, and the creation of awareness of the health risks of urban agriculture (de Zeeuw *et al.* 2000: 168-171). One of the contradictions of the legal setting of urban agriculture in Nakuru – and in Kenya in general and many other former British colonies as well – is that, according to the local by-laws, farming in town is an illegal activity, while the Municipality of Nakuru is just one of the extension divisions in Nakuru District for the Ministry of Agriculture. In other words, Nakuru Municipality has its own extension officers, for whom an urban farmer is as much a farmer as a rural farmer. Yet, the research project revealed that almost none of the Nakuru crop cultivators and a minority of the livestock keepers had been visited in 1998 by an extension officer (some other livestock keepers had received professional assistance from non-ministerial officers).[6] The study also showed, however, that professional assistance for livestock keepers did have a positive impact on rearing systems, the use of modern inputs and waste disposal. Among the recommendations formulated by the participants of the November 2002 workshop were issues such as educating urban farmers (by extension officers) on appropriate production methods, raising awareness about the environmental impact of farming in town, and space optimization (see Annex 11). Although to some extent, these issues were already being put in

---

[5] Personal communication, W.S. Wanyonyo, Environmental Officer, Municipal Council of Nakuru, May 2005.

[6] The numbers might have been higher if the study had not been restricted to the built-up area, i.e. if the peri-urban areas – in this study defined as the area between the built-up area and the municipal boundary – had been included.

practice by both the extension officers and ARDP officers, only a few urban farmers were actually being reached. A good example of practical research that benefits farmers with very tiny plots was the cultivation of crops in a large bag that requires very little inputs (see Photo 17) which was developed by ARDP and successfully introduced to some farmers. And ECLOF's micro-credit scheme, which started in Nakuru in 2001, signifies another positive development.

Photo 17 Experimenting with bag cultivation (*sukuma wiki*) in the ARDP compound. (Dick Foeken, 2001)

(3) Integration of urban agriculture in environmental policy implies the promotion of the safe re-use of urban organic waste and waste water by urban farmers, and the promotion of ecological farming methods (de Zeeuw *et al.* 2000: 171-173). Recycling urban organic waste for urban crop cultivation purposes is often mentioned as one of the major benefits of urban agriculture for the urban environment. Unfortunately, this is still a rather utopian picture in sub-Saharan Africa. Large-scale composting requires a well-functioning, municipality-wide collection system, installations for the composting process and a distribution system for the compost. This requires large start-up investments, money and technical skills for the operating and maintenance of the whole system, as well as organizational skills. A serious attempt – probably the first one in sub-Saharan Africa, with the possible exception of South Africa – to set up such a system was done in Cotonou, Benin, but it did not last long (Brock & Foeken 2005). Nakuru faces the same urban waste problems as almost all African urban centres: a very limited collecting system (in the town centre and high-income areas only), a dump where all non-separated waste is deposited, piles of garbage in the streets of the low-income areas and a chronic lack of means to keep the system running. Small-scale waste recycling offers better prospects in a town like Nakuru. There are many examples of successful waste recycling by small-scale enterprises and cooperatives in towns and cities of developing countries.[7] An environmental project in Nakuru, carried out by a partnership of the Municipal Council and local groups, whereby waste was collected in some low-income residential areas, was successful for some time (Mwangi 2003), but had been more or less abandoned by the time NUAP started.

The same applies to the use of waste water for urban crop cultivation. Although very successful systems operate in East Asia,[8] this is probably a distant dream for a town like Nakuru. The municipality does have a sewage system with two treatment plants, but the sewers cover only part of the town and the treatment plants do not function properly. One of the necessary improvements in the context of the development of urban agriculture in Nakuru, as mentioned during the workshop, concerns the rehabilitation of the sewage system, such that domestic waste is separated from industrial waste (the latter containing high concentrations of heavy metals found in the soils and plants irrigated with sewage water). This could be an important step to optimising benefits and minimising health risks from the use of waste water for farming in town (see van de Hoek 2002).

---

7 See, for example, www.waste.nl.

8 See *Urban Agriculture Magazine* no. 14 (July 2005) on Urban Aquatic Production.

Another matter that needs serious attention concerns the three rivers (Njoro, Makalia and Nderit) flowing through the town and into Lake Nakuru, with its fragile ecosystem. The water quality of these rivers has deteriorated considerably over the last decades due to high organic loads (faecal coliform), silt, heavy metals, oils and pesticides (Mkawale 2000). It is this same water that is used by urban farmers with riverbank plots to irrigate their crops (not to mention the people who use it as drinking water). Moreover, the use of chemicals by such farmers increases the pollution of these rivers, although most of the damage is likely to be caused further upstream by farms and industries there.

The promotion of ecological farming methods does take place in Nakuru. According to some of the local extension officers,[9] organic farming is encouraged among Nakuru producers. However, given the fact that virtually none of the Nakuru crop cultivators (in the built-up area) had been visited by an extension officer and that half of them used at least one chemical input, this 'policy' cannot yet be called successful. Organic farming is also encouraged by a local (Danish-sponsored) NGO called SENVINET (Strategic Environmental Network) in the context of an environmental-awareness programme aimed at school children. As we have seen in this book, school farming is common in Nakuru town and almost 30 schools, mainly primary schools, participate in the SENVINET programme.[10]

From the studies undertaken so far in Nakuru and from recent developments in the town, it has become clear that urban farming (1) is omnipresent in Nakuru and is likely to remain so in the future, (2) is a very important livelihood element for many households, (3) is tolerated by the local authorities, and (4) is receiving attention from policy-makers and NGOs. In short, these are all necessary conditions for the further development and improvement of this economic sector.

However, at least one important question remains: to what extent can urban farming in Nakuru be a tool in poverty reduction? Many Nakuru households are living below the 'absolute poverty line', but it is this group that appeared to be substantially under-represented among urban farmers because of a lack of access to land and capital. This is the group that would benefit most from urban farming, especially those who also lack access to rural land. But, as usual, it is the very poor who do not benefit from development policies and programmes. A way to help them could be to make certain tracts of land available, located

9 Personal communication, May 2005.

10 It is not possible to assess the success of the programme. A special sub-study in an upcoming research project on school farming in Nakuru town (with fieldwork envisaged in 2006) will be devoted to this issue.

either in or at short distance from the municipality and lease out small parcels of this land at a modest fee. In the words of Mushamba (2002: 9):[11]

> Most urban poor have little land of their own to produce food they need. However, most land in African municipalities is public and institutional, even private. This is land that planners at municipal level should consider as a resource available at any time for food production.

There are examples of municipalities (Dar es Salaam, Maputo) where open spaces have been made available for vegetable production and harvesting grass (Ibid.). To some extent, Morogoro town in Tanzania could also serve as an example (see Foeken *et al.* 2004: 125). There, the municipal authorities sub-divided former sisal estates, the plots to be used by people from each of the town's nineteen wards. In addition, an area of 3,000 hectares outside Morogoro was acquired by the local branch of the Ministry of Agriculture, for use by townspeople who would otherwise have had no access to land. However, whether it will be the very poor of the town that will benefit from such developments remains to be seen.

---

[11] Shingirayi Mushamba is Senior Programme Officer and Urban Agriculture Programme Coordinator at MDP (Municipal Development Partnership for Eastern and Southern Africa: A Partnership Enabling Local Government Capacity) in Harare and presented the keynote address at the NUAP workshop on 27 November 2002.

Annexes

Tables

Annex 1: The Nakuru Urban Agriculture Research Project (NUAP)

The Nakuru Urban Agriculture Research Project consists of the following studies:

1) General survey among 594 households (fieldwork: 1999) and additional interviews with 30 farming households selected from the study population of the general survey (2000). Researchers: Dick Foeken (ASC), Samuel Ouma Owuor (University of Nairobi).

   - Foeken, D. & S.O. Owuor (2000), *Urban farmers in Nakuru, Kenya.* Leiden: African Studies Centre, ASC Working Paper no. 45 (also on www.cityfarmer.org/nakuru.html).
   - Foeken, D. & S.O. Owuor (2000), Livestock in a middle-sized East-African town: Nakuru. *Urban Agriculture Magazine* 1(2): 20-22.
   - Foeken, D. & S.O. Owuor (2001), Multi-spatial livelihoods in Sub-Saharan Africa: Rural farming by urban households – The case of Nakuru town, Kenya. In M. de Bruijn, R. van Dijk & D. Foeken, eds., *Mobile Africa: Changing patterns of movement in Africa and beyond*, pp. 125-140. Leiden: Brill.
   - Foeken, D., S.O. Owuor & W. Klaver (2002), *Crop cultivation in Nakuru town, Kenya: Practice and potential.* Leiden: African Studies Centre, ASC Working Paper no. 50 (also on http://www.ascleiden.nl/pdf/workingpaper50.pdf).
   - Owuor, S.O. (2005), Coping with urban poverty: A study of farming within Nakuru town, Kenya. *Hekima – Journal of the Humanities and Social Sciences* 3(1): 84-101.
   - Owuor, S.O. & D. Foeken (2006), Surviving in the neighbourhoods of Nakuru town, Kenya. In P. Konings & D. Foeken, eds., *Crisis and creativity. Exploring the wealth of the African neighbourhood*, pp. 22-45. Leiden: Brill.

2) Impact of urban farming on the food and nutritional situation of the households involved (fieldwork: 2000). Researchers: Wijnand Klaver (ASC), Dick Foeken (ASC), Samuel Ouma Owuor (University of Nairobi).

   - Results integrated in this book.

3) Environmental aspects of farming in Nakuru town (fieldwork: 2000). Researcher: Ernest Oyieko Nyandwaro (Kenyatta University).

   - Nyandwaro, E.O. (2006), *Environmental impact of urban farming: A case study of Nakuru town, Kenya.* Nairobi: Kenyatta University, School of Pure and Applied Sciences, MSc thesis (version submitted for examination).

4) Impact of support for urban farmers on the income, food and nutritional situation of the households involved (fieldwork: 2000-01). Researcher: Peter Wambugu King'ori (University of Nairobi).

   - King'ori, P.W. (2006), *Food security, child nutritional status and incomes of urban farming households in Nakuru town, Kenya: A comparative study.* Nairobi: University of Nairobi, Applied Nutrition Programme, MSc thesis (version submitted for examination).

5) School farming in Nakuru town (fieldwork: 2000/01). Researcher: Elizabeth Correta Odera (University of Nairobi).

- Odera, E.C. (2006), *The implications of urban school farming on food security and nutrition: A study in Nakuru town.* Nairobi: University of Nairobi, Applied Nutrition Programme, MSc thesis (version submitted for examination).

6) Decision-making around farming in town (fieldwork: 2001). Researcher: Nicole Versleijen (Wageningen University).

- Versleijen, N. (2002), *Sukuma! A social analysis of urban agriculture: Case studies from Nakuru Town, Kenya.* Wageningen: Wageningen University and Research Center, Department of Rural Development Sociology, MSc thesis.

7) Legal aspects of urban farming in Nakuru town (1998-2000/2005, interviews and literature study). Researcher: Dick Foeken (ASC).

- Foeken, D. (2005), *Urban agriculture in East Africa as a tool for poverty reduction: A legal and policy dilemma?* Leiden: African Studies Centre, ASC Working Paper no. 65 (also on http://www.ascleiden.nl/pdf/workingpaper65.pdf)
- Foeken, D. (2006), Legislation, policies and the practice of urban farming in Nakuru: Contradictions abound. *Urban Agriculture Magazine* 16 (July 2006).
- Foeken, D. (fc), Urban agriculture and the urban poor in East Africa: Does policy matter? (to be published in *African Dynamics*, vol. 7)

8) Rural farming by urban households: The case of Nakuru town (fieldwork: 2002-04). Researcher: Samuel Ouma Owuor (University of Nairobi).

- Owuor, S.O. (2002), Enhancing food security in African cities: Rural farming by urban households – The case of Nakuru town, Kenya. Paper presented at the "Workshop on urban policy implications of enhancing food security in African cities", Nairobi, 27-31 May 2002.
- Owuor, S.O. (2003), *Rural livelihood sources for urban households. A study of Nakuru town, Kenya.* Leiden: African Studies Centre, ASC Working Paper 51.
- Owuor, S.O. (2004), Urban households ruralizing their livelihoods: The changing nature of urban-rural linkages in an East African town. A paper presented at the "African Studies Centre Seminar Series", Leiden, Netherlands, 16 December 2004.
- Owuor, S.O. (2006), *Bridging the urban-rural divide: Multi-spatial livelihoods in Nakuru town, Kenya.* Leiden: African Studies Centre, Research Report 81.

9) Economic aspects of urban farming in Nakuru town (fieldwork in progress). Researcher: Reuben Anyango (Egerton University).

10) School farming and school feeding in Nakuru town, Kenya (fieldwork planned for 2006). Researchers: Alice Mboganie Mwangi (University of Nairobi), Dick Foeken (ASC), Samuel Ouma Owuor (University of Nairobi), Wijnand Klaver (ASC).

Annex 2: Sampling and the study population

The 1999 survey

To obtain an overall view of urban agriculture in Nakuru town, a general survey was carried out in June-July 1999. To ensure a representative sample of 600 households from among Nakuru's population, the 15 clusters of the Kenyan Central Bureau of Statistics (CBS) were used (see Map A2.1). Together, the clusters counted about 1,400 households. Therefore, from each cluster a 43% (600 out of 1400) random sample was drawn (Table A2.1). In the end, 594 households were covered.

It is important to note that the 15 CBS clusters cover the *built-up area* of Nakuru Municipality. There are no clusters in the peri-urban areas i.e. between the (more or less densely) built-up area and the boundary of the municipality. The latter are former rural areas that were incorporated into the municipality after the boundary extensions of 1972 and 1992 and still have a predominantly agricultural character: the area of large-farms in the west, small farms in the southwest and medium-sized farms in the far east. The 'white square' in the far east is a military area (the Lanet Army Barracks). In short, we are dealing with urban farming in the strict sense in this book (see Chapter 1) by *excluding* the peri-urban areas.

Several problems were encountered during the fieldwork, the major ones being:

- Outright refusal by some households to respond to our questions. This was mainly due to the fact that by the time of the survey, there was another general survey on HIV/Aids being carried out in the same clusters. Due to the sensitive nature of those questions, most households, especially the ones in high-density areas, were suffering from 'questionnaire fatigue'. They were, therefore, not ready to take any more questions from us, thinking that we had the same types of questions. This at one time led to the households playing a 'hide and seek' game with the assistants, especially in one of the clusters in Kwa Rhonda. Such refusals were solved by 'replacing' these households, but only after more than three attempts at persuasion.
- Almost all the houses in one of the other clusters in Kwa Rhonda were vacant due to inter- and intra-estate mobility. Apparently, the landlord had increased monthly rents the month before the survey. Most tenants had, therefore, moved out to affordable dwellings in the neighbourhood not known to the assistants. This problem was solved by selecting another structure within or just outside the CBS cluster to replace the vacant houses or by selecting another household within the cluster to be used as a 'replacement'.
- A number of 'call-backs' for household heads who were working during normal working hours in the week forced assistants to go back to the households late in the evening and at weekends. This problem occurred particularly in the low-density/-

Map A2.1 Distribution of the research clusters

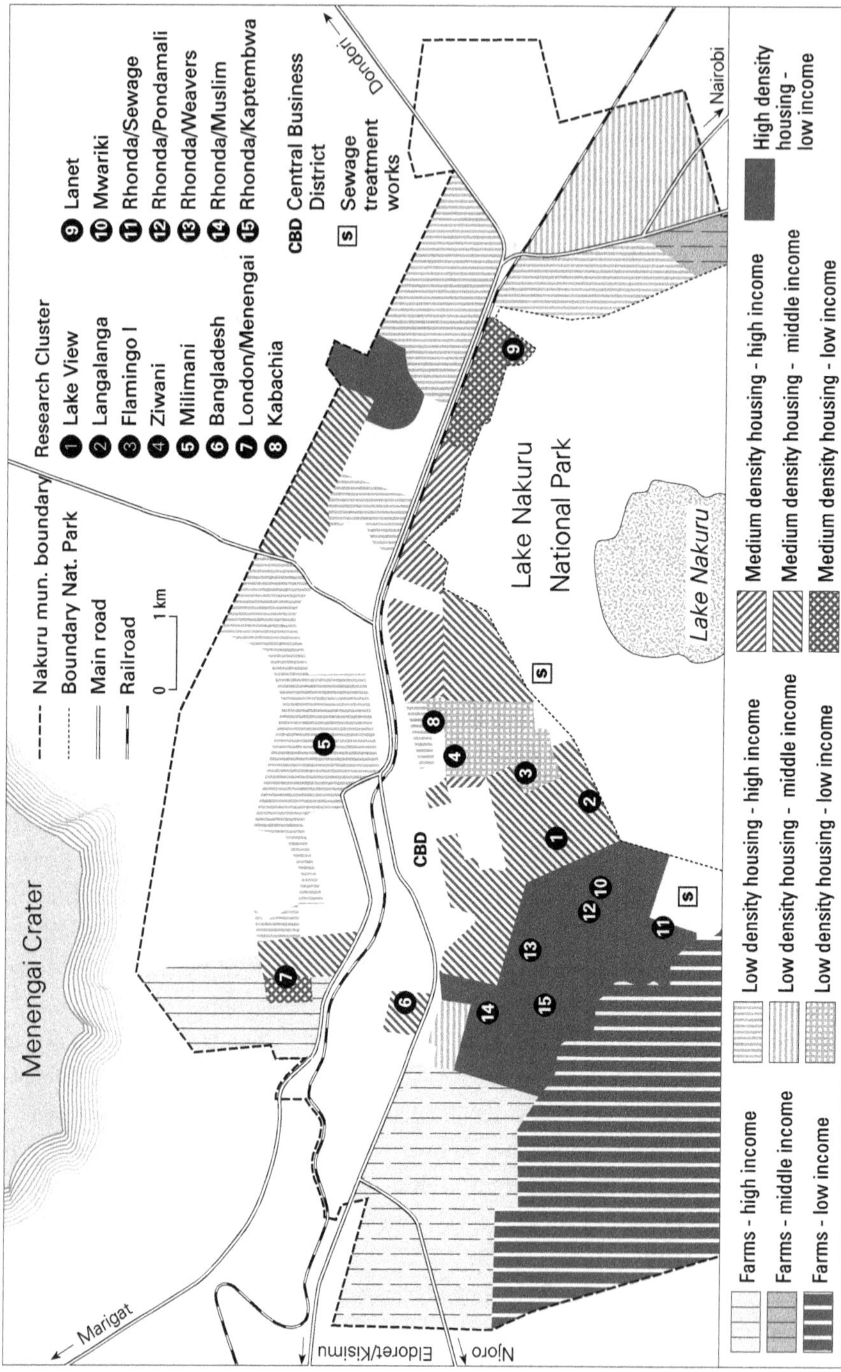

high-income cluster of Milimani. It was dealt with by making appointments (call-backs) at the convenience of the household head. If there were more than three call-backs the household was replaced.

- In some low and medium-density clusters, there was a problem of suspicion that in turn led to reluctance by some respondents to cooperate. This problem was overcome by giving a careful introduction and details about the purpose of the research and asking whether the respondent had any questions, concerns or reservations about the survey. The CBS enumerators were very helpful in this since most of the households knew them personally.

Table A2.1 Household samples

	cluster name	estate name	housing density*	no. of h'holds drawn	no. of h'holds covered in 1999	no. of h'holds covered in 2000
1	Lake View	Lake View	medium	43	42	10
2	Langalanga	Langalanga	medium	45	43	7
3	Flamingo I	Flamingo I	low	37	37	10
4	Ziwani	Ziwani	low	42	43	8
5	Milimani	Milimani	low	9	8	4
6	Bangladesh	Bangladesh	medium	39	39	6
7	London/Menengai	Menengai	medium	26	28	10
8	Kabachia	Kabachia	low	42	39	24
9	Lanet	Lanet	medium	56	56	14
10	Mwariki	Mwariki	high	56	55	8
11	Rhonda/Sewage	Kwa Rhonda	medium	25	23	6
12	Rhonda/Pondamali	Kwa Rhonda	high	46	48	7
13	Rhonda/Weavers	Kwa Rhonda	high	39	38	5
14	Rhonda/Muslim	Kwa Rhonda	high	34	33	12
15	Rhonda/Kaptembwa	Kwa Rhonda	high	61	62	5
	Total			600	594	136

* Based on own observations in the field. The main criterion used is the amount of space that can potentially be used for agricultural purposes relative to the total surface of the residential plots in a certain cluster. Hence, the densities in this table do not fully overlap with the ones given by MCN 2000, Figure 3.6 (see Map 2.4).

The 2000 survey

A second survey was held in September-October 2000, specifically focusing on aspects of food consumption and nutrition. From the outset, the intention was to select 150 urban-farming households and 50 'non-farmers', the former group being much larger than the latter to allow for further sub-analyses. The study population of 2000 was *not* therefore a representative sample of Nakuru's population (unlike the 1999 study population) because of its focus on *comparison* between the two groups. Households were to be selected from the 1999 study population. However, urban-farming households were only eligible if they met certain minimum conditions, i.e. a plot size of at least 10 m^2 (crop cultivators) and/or at least one cow or five goats/sheep or ten small livestock (livestock keepers). A second condition was that the household had been involved in urban farming in both 1998 and 1999. The idea behind introducing these conditions was to make comparisons between farmers and non-farmers more worthwhile. At the time of sampling, there appeared to be 153 eligible urban-farming households; so it was decided to use them all. As for the 50 non-farming households, these were intended to be selected in a stratified way as follows: 57% of them from non-farming households *with* under-fives (children younger than 60 months: one in three non-farming households) and 43% of them from non-farming households *without* under-fives (one in eight non farming households). This was done partly to have sufficient households with young children, and also not to have *only* young households, which would make it impossible to compare the farming and non-farming groups. To allow for drop-outs, a larger sample of 80 non-farming households was selected. During the fieldwork, major problems arose as many households appeared to have moved or refused to cooperate. Upon verification of recent urban farming activities, about 10% of the urban-farming households appeared to be misclassified: there were 137 (instead of 153) eligible urban-farming households and 96 (instead of 80) eligible non-farming households Since no replacement households were available in the urban-farming group and since only a limited number (17) could be realized in the non-farming group, only 74 farming households and 62 non-farming households could be covered in the end. The rate of 'dropouts' for various reasons in the two groups was comparable. About 40% of the target number was not attained even after replacements in the non-farming group, but at cluster level the rates showed marked differences, ranging from 0% in Flamingo, about 25% in three clusters (Ziwani, Kabachia and Rhonda Muslim) to 67% in Rhonda Weavers.

The respondents in the 1999 and 2000 surveys

Table A2.2 shows various characteristics of the respondents in the surveys carried out in 1999 and 2000. The large majority were either the head of the household him/herself or the male-head's spouse (the spouse was always a woman). In some cases, another person was interviewed, usually a brother or sister of the household head or a grown-up child. The majority of the respondents were between 20 and 40 years of age.

Table A2.2 Characteristics of respondents (%)

	(N=)	1999 (594)	2000 (135)
relation to household head	household head	40.2	27.2
	spouse	46.6	54.4
	brother/sister	5.4	2.9
	child (adult)	6.7	14.7
	other	1.0	0.7
	Total	100	100
sex	male	30.6	25.7
	female	69.4	74.3
	Total	100	100
age (years)	<20	15.5	4.4
	20-29	39.9	37.0
	30-39	25.1	33.3
	40-49	11.4	12.6
	50+	8.1	12.6
	Total	100	100

Source: 1999 and 2000 surveys.

The in-depth survey

An in-depth survey was also held among 30 urban-farming households in 2000. These all met the same conditions as the urban-farming households in the previous survey. Since the study paid specific attention to low-income households, an additional condition was introduced, namely that only households from income classes 1 and 2 (i.e. with a monthly cash income of less than Ksh. 10,000) be included. There appeared to be 74 households meeting all these criteria: 39 crop cultivators (10 randomly selected), 9 livestock keepers (all selected) and 25 mixed farmers (11 randomly selected). During the fieldwork, six households appeared to have moved or refused to participate. These were replaced by another household in the same category or, in the case of 'pure' livestock keepers, by their nearest neighbour meeting the same criteria. The households were approached with a semi-structured questionnaire focusing on urban-farming activities and most were visited twice.

Additional household information

Additional, qualitative information presented in this book was obtained from a number of sources. The first was the Masters thesis by Nicole Versleijen (2002), which is an anthropological study with detailed information on five urban-farming households. Results from the Masters theses by Ernest Nyandwaro and by Correta Odera constitute the bulk of Chapters 8 and 9, respectively, while findings from the Masters thesis carried out by Peter King'ori are incorporated in Chapter 7. Another source concerns some case studies in Samuel Owuor's PhD study. The free use granted by the authors of these sources is greatly appreciated.

Table A2.3 Some basic characteristics of the sampled households by survey (%)

	(N=)	1999 (594)	2000 (136)
household size (members)	1	11.3	-.-
	2-4	53.7	34.6
	5-7	27.9	44.1
	8+	7.1	21.3
	Total	100	100
monthly cash income (Ksh.)	up to 5,000	53.2	22.1
	5,001-10,000	28.6	33.8
	10,001-20,000	12.7	28.7
	>20,000	5.5	15.4
	Total	100	100
sex of household head	male	80.3	89.7
	female	19.7	10.3
	Total	100	100
age of household head	<20	2.3	-.-
	20-29	29.3	10.3
	30-39	35.0	34.6
	40-49	20.1	30.1
	50-59	10.4	19.9
	60+	2.9	5.1
	Total	100	100

Source: 1999 and 2000 surveys.

Annex 3: The farmers

Table A3.1 Farmers, crop cultivators and livestock keepers in town by cluster, 1998 (%)

	cluster name	housing density	households in sample (=N)	% urban farmers	% urban crop cultivators	% urban livestock keepers
1	Lake View	medium	42	28.6	26.2	7.1
2	Langalanga	medium	43	32.6	25.6	11.6
3	Flamingo I	low	37	29.7	16.2	18.9
4	Ziwani	low	43	25.6	16.3	16.3
5	Milimani	low	8	87.5	87.5	75.0
6	Bangladesh	medium	39	38.5	25.6	30.8
7	London/Menengai	medium	28	71.4	67.9	46.4
8	Kabachia	low	39	82.1	82.1	35.9
9	Lanet	medium	56	37.5	33.9	17.9
10	Mwariki	high	55	27.3	10.9	21.8
11	Rhonda/Sewage	medium	23	39.1	34.8	26.1
12	Rhonda/Pondamali	high	48	31.3	14.6	18.8
13	Rhonda/Weavers	high	38	31.6	23.7	21.1
14	Rhonda/Muslim	high	33	15.2	15.2	-.-
15	Rhonda/Kaptembwa	high	62	16.1	4.8	14.5
	Total		594	35.2	26.9	20.4

Source: 1999 survey.

Table A3.2 Urban farmers and non-farmers: household characteristics (%)

		urban farmers (N=209)	non-farmers (N=385)
household size (number of members)	1	5.3	14.5
	2-4	37.3	62.6
	5-7	44.5	19.0
	8 or more	12.9	3.9
	Total	100	100
household income class (Ksh./month)	up to 5,000	33.2	64.0
	5,001 - 10,000	36.1	24.6
	10,001 - 20,000	21.0	8.2
	more than 20,000	9.8	3.2
	Total	100	100
house ownership	own house	21.5	4.9
	rent the house	78.0	94.5
	plot keeper	0.5	0.5
	Total	100	100
housing density of estate	high	27.3	46.5
	medium	69.4	53.2
	low	3.3	0.3
	Total	100	100

Source: 1999 survey.

Table A3.3 Urban farmers and non-farmers: characteristics of household heads (%)

		urban farmers (N=209)	non-farmers (N=385)
age (years)	less than 20	1.4	2.7
	20-29	14.5	37.6
	30-39	30.9	37.3
	40-49	30.0	14.6
	50-59	18.4	5.9
	60 or more	4.8	1.9
	Total	100	100
sex	male	84.7	77.9
	female	15.3	22.1
	Total	100	100
ethnic group	Kikuyu	38.8	46.9
	Luo	26.3	16.4
	Luhya	12.9	20.1
	Kalenjin	6.7	5.2
	Kisii	7.7	4.2
	Kamba	3.8	3.9
	Asian	1.9	1.3
	others*	1.9	2.0
	Total	100	100
type of residence	full-time resident	92.3	97.4
	regularly absent	7.7	2.6
	Total	100	100
marital status	single	10.6	22.9
	married monogamously	73.6	68.8
	married polygamously	8.7	1.8
	divorced/separated/widowed	7.2	6.5
	Total	100	100
educational level	none	1.4	3.1
	primary school, up to Standard 4	6.7	5.0
	primary school, Standard 5-8	23.9	33.9
	secondary school	45.0	47.0
	more than secondary school	23.0	11.0
	Total	100	100
occupational status	regularly employed	40.2	41.3
	temporarily employed	3.3	5.2
	self-employed	45.5	39.7
	casual labourer	7.2	9.4
	housewife	1.0	1.6
	unemployed/not employed	2.9	2.8
	Total	100	100

* Meru (0.5% and 0.8%), Somali (0% and 0.5%), Teso (0 and 0.3%), Maasai (0.5% and 0%), Taita (0% and 0.3%), Mijikenda (0% and 0.3%) and Pare (0.5% and 0%).

Source: 1999 survey.

Table A3.4 Urban farmers and non-farmers: migration history of household heads (%)

		urban farmers (N=209)	non-farmers (N=385)
migration status	born in Nakuru town	13.0	13.4
	immigrant from within Kenya	85.1	85.9
	immigrants from outside Kenya	1.9	0.8
	Total	100	100
province of origin (immigrants from within Kenya only)	Nairobi	1.4	2.7
	Central	22.6	22.9
	Rift Valley	21.5	27.1
	Western	13.6	22.3
	Nyanza	34.5	19.5
	Eastern	5.6	4.9
	North Eastern	-.-	0.3
	Coast	-.-	0.9
	Total	100	100
year to come to Nakuru (immigrants only)	before 1970	9.0	2.8
	1970 - 1979	28.1	8.0
	1980 - 1989	34.3	29.0
	1990 - 1994	20.2	27.8
	1995 - 1999	8.4	32.4
	Total	100	100
main reason to come to Nakuru (immigrants only)	lack of land in home area	1.7	0.6
	lack of work in home area	0.6	0.6
	to look for work / to work	79.9	75.7
	had relatives in Nakuru	7.3	7.0
	followed spouse/came with parents	7.2	10.1
	schooling/training	1.1	3.3
	tribal clashes in home area	1.7	2.4
	had a plot in Nakuru	0.5	-.-
	health reasons	-.-	0.3
	Total	100	100

Source: 1999 survey.

Table A3.5 Non-farmers: reasons for not farming in town by type of farming (%)

		no crop cultivation (N=434)		no livestock keeping (N=473)	
		reasons (>100%)	main reason	reasons (>100%)	main reason
land issues	no access to urban land	85.7	75.6	74.4	62.8
	plot too far away	0.2	-.-	0.2	0.2
	land used by someone else	0.2	0.2	-.-	-.-
	have access to rural land	8.1	2.3	5.7	1.9
lack of other resources	no capital	28.6	9.4	24.1	10.6
	lack of time	7.1	4.4	6.6	4.4
	no labour available	1.6	0.5	2.7	1.1
legal consider-ations	harassment	1.8	0.2	4.7	1.7
	not allowed in town	0.5	0.5	0.2	0.2
	disapprove of it	3.5	0.9	6.3	2.5
	landlord disapproves	1.8	-.-	5.1	1.3
	neighbours complain	0.5	-.-	0.4	0.4
	theft/insecurity	0.9	0.2	2.3	1.9
	tribal clashes	0.2	0.2	0.2	0.2
other reasons	not worthwile	5.3	2.5	7.2	5.1
	had not thought about it	0.9	0.7	2.5	2.5
	crop failure in the past	0.5	0.2	-.-	-.-
	destruction by animals	0.9	-.-	0.6	0.4
	diseases	-.-	-.-	1.5	1.1
	was not in Nakuru	2.3	2.1	1.7	1.7
	Total		100		100

Source: 1999 survey.

Table A3.6 Rural farmers and non-farmers: household characteristics (%)

		rural farmers (N=366)	non-farmers (N=228)
household size (number of members)	1	10.4	12.7
	2-4	54.4	52.6
	5-7	28.1	27.6
	8 or more	7.1	7.0
	Total	100	100
household income class (Ksh./month)	up to 5,000	47.8	61.8
	5,001 - 10,000	32.1	23.1
	10,001 - 20,000	14.2	10.2
	more than 20,000	5.9	4.9
	Total	100	100
population density of estate	high	67.8	58.3
	medium	30.6	40.8
	low	1.6	0.9
	Total	100	100

Source: 1999 survey.

Table A3.7 Rural farmers and non-farmers: characteristics of household heads (%)

		rural farmers (N=366)	non-farmers (N=228)
age (years)	less than 20	2.0	4.5
	20-29	31.6	25.1
	30-39	32.7	38.1
	40-49	21.8	17.0
	50-59	10.6	9.9
	60 or more	1.4	5.4
	Total	100	100
sex	male	85.2	74.4
	female	14.8	27.6
	Total	100	100
ethnic group	Kikuyu	36.2	56.6
	Luo	24.4	12.7
	Luhya	20.8	12.3
	Kalenjin	7.7	2.6
	Kisii	5.5	5.3
	Kamba	4.1	3.5
	Asian	-.-	3.9
	others	1.4	3.1
	Total	100	100
type of residence	full-time resident	95.6	95.6
	regularly absent	4.4	4.4
	Total	100	100
marital status	single	14.8	24.6
	married monogamously	76.4	61.0
	married polygamously	4.9	3.1
	divorced/separated/widowed	3.8	11.4
	Total	100	100
educational level	none	0.8	5.3
	primary school, up to Standard 4	4.1	7.9
	primary school, Standard 5-8	31.0	29.4
	secondary school	48.1	43.4
	more than secondary school	15.9	14.0
	Total	100	100
occupational status	regularly employed	46.3	32.2
	temporarily employed	4.4	4.8
	self-employed	38.6	46.7
	casual labourer	7.7	10.1
	housewife	0.8	2.2
	unemployed/not employed	2.1	4.0
	Total	100	100

Source: 1999 survey.

Annex 4: The crops

Table A4.1 Characteristics of urban plots by housing density (%)

	(N=)	high density (35)	medium density (138)	low density (7)	total (180)
location	riverside	-.-	0.7	-.-	0.6
	roadside	2.9	4.5	-.-	4.0
	railwayside	-.-	10.4	-.-	8.0
	under powerline	-.-	2.2	-.-	1.7
	in own compound	52.9	60.4	100.0	60.6
	within estate	29.4	14.2	-.-	16.6
	in other estate	11.8	5.2	-.-	6.3
	elsewhere	2.9	2.2	-.-	2.3
	Total	100	100	100	100
distance to plot in minutes walking	less than 10 minutes	65.7	69.9	100.0	70.0
	10-30 minutes	11.4	13.0	-.-	12.2
	30-60 minutes	11.4	10.1	-.-	10.0
	more than 60 minutes	11.4	7.2	-.-	7.8
	Total	100	100	100	100
plot size in m^2	<10	11.4	14.5	-.-	13.4
	10-49	2.9	21.0	16.7	17.3
	50-99	8.6	9.4	-.-	8.9
	100-499	28.6	18.8	-.-	20.1
	500-999	11.4	8.0	-.-	8.4
	1000-4999	34.3	22.5	83.3	26.8
	5000+	2.9	5.8	-.-	5.0
	Total	100	100	100	100
ownership of plot	own land	57.1	25.5	57.1	33.0
	landlord	28.6	50.4	42.9	45.8
	government	2.9	17.5	-.-	14.0
	other	11.4	6.6	-.-	7.3
	Total	100	100	100	100

Source: 1999 survey.

Table A4.2 Crops cultivated in Nakuru (% of plots; N=180)

maize	61.7	pawpaw	6.1	(*mbiriganya*)	1.7
kales	61.1	pumpkins	5.6	*mavaki* (local	
beans	56.1	green peas	5.0	vegetable)	1.7
onions	25.0	avocado	3.9	millet	1.7
spinach	20.0	Napier grass	3.9	passion fruit	1.7
tomatoes	19.4	cassava	3.3	American herb	1.1
Irish potatoes	17.8	amaranth (*terere)*	3.3	mango	1.1
bananas	15.6	black night shade		capsicum	0.6
cowpeas	15.6	(*managu)*	2.8	guava	0.6
spider plant		cucumber	2.8	lettuce	0.6
(*saget)*	11.1	parsley (*dhania)*	2.8	loquarts	0.6
pepper	7.2	oranges	2.2	mushrooms	0.6
sugarcane	7.2	sweet potatoes	2.2	pineapple	0.6
cabbage	6.1	arrow root	1.7	strawberry	0.6
carrots	6.1	egg plant		water melon	0.6

Source: 1999 survey.

Table A4.3 Characteristics of urban crop cultivation by person responsible

		male head (N=48)	wife of male head (N=88)	female head (N=23)
% households cultivating:	maize	72.9	58.0	47.5
	beans	70.8	51.1	39.1
	kales	54.2	61.4	65.2
	spinach	10.4	23.9	21.7
	onions	10.4	29.5	30.4
	tomatoes	14.6	21.6	17.4
	cowpeas	12.5	15.9	17.4
	bananas	18.8	10.2	17.4
	Irish potatoes	10.4	15.9	8.7
	saget	4.2	11.4	13.0
% households using:				
- traditional inputs[a]	zero	14.6	12.5	26.1
	one	33.3	39.8	26.1
	two to three	52.1	47.7	47.8
- modern inputs[b]	zero	22.9	22.7	56.5
	one	29.2	37.5	39.1
	two to four	47.9	39.8	4.3
- chemical inputs[c]	zero	37.5	50.0	91.3
	one	27.1	29.5	4.3
	two to three	35.5	20.4	4.3
- irrigation	yes	18.8	48.9	52.2

Notes: a) Manure, crop residues and local seeds.
b) Chemical fertiliser, pesticides, insecticides and improved seeds.
c) Chemical fertiliser, pesticides and insecticides.

Source: 1999 survey.

Table A4.4 Problems with crop cultivation in town (%; N=160)

type of problem	mentioned as a problem	mentioned as the main problem
no problem	16.3	16.3
theft of crops	36.6	24.4
inadequate rainfall	35.0	24.4
destruction by animals	23.8	10.0
pests/insects	22.5	8.8
lack of water for irrigation	12.5	9.4
diseases	9.4	2.5
lack of inputs/capital	6.9	1.9
harassment	2.5	-.-
bad quality seeds	1.9	-.-
poor soil	1.3	-.-
lack of space/land	1.3	1.3
lack of labour	0.6	0.6
weeds	0.6	0.6
too much rainfall	0.6	-.-
poor seasonal timing	0.6	-.-
sewage burst	0.6	-.-
Total		100

Source: 1999 survey.

Annex 5: The animals

Table A5.1 Problems with livestock keeping by type of livestock (%)*

(N=; households)	large livestock (33)	small livestock (108)	all households (121)
A) Mentioned as problem			
- no problem	9.1	11.1	10.7
- diseases	75.8	71.3	71.9
- theft	24.2	20.4	21.5
- lack of feed	27.3	12.0	14.9
- lack of funds/capital	15.2	10.2	10.7
- lack of safe drinking water	24.2	6.5	9.1
- predators	3.0	10.2	9.1
- lack of space	3.3	7.4	6.6
- harassment	12.1	2.8	5.8
- lack of market	6.1	3.7	3.3
- pests/parasites	3.0	1.9	2.5
- lack of labour	6.1	0.9	1.7
- nuisance	3.0	0.9	1.7
B) Mentioned as main problem			
- no problem	9.1	11.1	10.7
- diseases	45.5	61.1	57.0
- theft	9.1	7.4	8.3
- lack of feed	12.1	1.9	3.3
- lack of funds/capital	6.1	6.5	6.6
- lack of safe drinking water	9.1	1.9	3.3
- predators	-.-	3.7	3.3
- lack of space	3.0	3.7	3.3
- harassment	3.0	-.-	0.8
- lack of market	-.-	0.9	0.8
- pests/parasites	-.-	1.9	1.7
- lack of labour	3.0	-.-	0.8
- nuisance	-.-	-.-	-.-
Total	100	100	100

* There are 20 households with both large and small livestock, hence the N of the column with 'all households' is 121 instead of 141.

Source: 1999 survey.

Annex 6: The benefits

Table A6.1 Crops: amounts self-consumed by crop type (N)

	amount self-consumed:	all	most	about half	less than half	small portion	none	total	calculation (%)
1	kale	48	34	10	8	6	3	109	75
2	maize	45	35	11	2	4	4	101	77
3	beans	51	22	7	3	2	3	94	77
4	onions	22	15	2	2	2	2	45	78
5	spinach	11	12	2	3	5	3	36	62
6	tomatoes	20	8	2	1	-	4	35	78
7	Irish potatoes	19	9	1	-	1	2	32	82
8	cowpeas	16	3	2	-	3	4	28	70
9	bananas	13	3	3	-	-	8	27	62
10	*saget*	8	5	1	2	1	2	19	68

Note: The amounts self-consumed (%) were calculated by translating the qualitative values of the amounts self-consumed into percentages as follows: all = 100%, most = 75%, about half = 50%, less than half = 30%, small portion = 10%, and none = 0%. N.B.: The percentages are at best indications only.
Source: 1999 survey.

Table A6.2 Calculation of energy from urban crop production

		no. of h'holds[a]	kg. harvested[a]	aggregate kgs	kcal/kg as purchased[b]	aggregate kcal.
1	kale	109	84	9,156	384	3,515,904
2	maize	101	224	22,624	3,630	82,125,120[c]
3	beans	94	75	7,050	3,390	23,899,500
4	onions	45	26	1,170	187.2	219,024
5	spinach	69	92	6,348	384	2,437,632
6	tomatoes	35	15	525	196	102,900
7	Irish potatoes	32	88	2,816	637.5	1,795,200
8	cowpeas	28	67	1,876	3,400	6,378,400
9	bananas	27	4	108	777.2	83,916
10	saget	19	33	627	224	140,448
11	other crops	d	32.7	4,184	2,308	9,656,672
12	total energy produced (1-11)	160		(56,484)		130,354,716
13	daily energy requirement per capita					2,200
14	annual energy requirement per capita (13x365 days)					803,000
15	annual energy requirements per household (14x3.5 persons/household)[e]					2,810,500
16	annual energy requirements 160 households (15x160 households)					449,680,000
17	annual energy requirements 594 households (15x594 households)					1,669,437,000
18	contribution of urban crop production to energy requirements 160 households (12/16x100%)					29.0%
19	contribution of urban crop production to energy requirements 594 households (12/17x100%)					7.8%

Notes:
a. From Table 4.4.
b. See Platt 1962.
c. Assuming that weight figures refer to dry mature kernels (100% edible).
d. From Annex 4, Table A4.1. As other crops include 31 items, the aggregate number of households is not meaningful.
e. Household size of 3.5 calculated as the total population in 1999 (239,000) divided by the total number of households (68,436). See Kenya 2000.

Source: 1999 survey.

Table A6.3 Urban farmers and non-farmers: general food security issues, 1998 and 1999 (%)

		1998		1999	
		urban farmers (N=209)	non-farmers (N=385)	urban farmers (N=71)	non-farmers (N=62)
always food enough last year?	yes, always	93.3	84.9	63.4	69.4
	most of the time	3.3	7.0	22.5	24.2
	about half of the time	1.9	3.4	12.7	1.6
	now and then	1.0	4.7	1.4	4.8
	never	0.5	-.-	-.-	-.-
	Total	100	100	100	100
most important food source last year	own urban production	5.3	-.-	1.4	-.-
	urban + rural production	9.6	-.-	-.-	-.-
	urban production + purchased	30.1	-.-	28.8	-.-
	purchased	36.8	68.1	61.6	90.3
	rural production	0.5	1.3	-.-	-.-
	rural production + purchased	17.7	29.9	8.2	9.7
	donations + purchased	-.-	0.8	-.-	-.-
	Total	100	100	100	100

Source: 1999 and 2000 surveys.

Table A6.4 Urban farmers and non-farmers: energy intake by food group

		Percentage households consuming food group			Mean energy intake (kcal/CU/day) by food group			Percentage contribution of each food group to total daily energy intake		
Food group	(N=)	UF (74)	non-UF (61)	Total (135)	UF (74)	non-UF (61)	Total (135)	UF (74)	non-UF (61)	Total (135)
Cereal products		100	98.4	99.3	1609	1604	1607	56.3	56.9	56.6
Grain legumes		47.3	39.3	43.7	154	114	136	5.3	4.6	5.0
Roots, tubers & starchy staples		67.6	75.4	71.1	126	139	132	5.0	5.6	5.3
Vegetables		100	100	100	146	144	145	5.9	5.6	5.8
Fruits		10.8	3.3	7.4	9	1	5	0.3	0.0	0.2
Products of animal origin		94.6	98.4	96.3	322	263	295	10.8	10.1	10.5
Seeds & nuts		2.7	1.6	2.2	5	0	3	0.2	0.0	0.1
Oils, fats & margarine		98.6	98.4	98.5	267	292	278	9.4	9.7	9.5
Miscellaneous		97.3	91.8	94.8	194	198	196	6.8	7.4	7.1
Total					2832	2755	2797	100	100	100

Note: UF = urban farmer.
Source: 2000 survey.

Note on consumer units and calculations: see next page.

Note on consumer units and calculations (Table A6.4)

To ensure a meaningful comparison between households, food intake data were related to household size, not simply measured on a per capita basis but in terms of the number of "consumer units" (or "adult equivalents"). One consumer unit corresponds to the daily energy requirements of a young adult male. The energy requirements of an individual (according to his/her sex, age, physiological condition and activity pattern) are expressed as a ratio of this unit. The sum of the individual ratios in the household thus represents a "physiological" household size in terms of "consumer units" (for a detailed explanation, see Klaver & Mwadime 1998). In the food consumption questionnaire, meal participation was recorded for six categories of household members: children under three years of age, children 3-15 years of age, mother, father, others and visitors. These were assigned average consumer units as follows: 0.4, 0.7, 0.8, 1.0, 0.9 and 0.8, respectively. For each dish, the numbers (in consumer units) of household members not reported to have been eating elsewhere for that meal, and visitors, were added up. The amount of food in each dish consumed was divided by the number of consumer units. The idea is that a person skipping a meal but not eating elsewhere is still counted among those whose food requirements should be met by that household consumption. The amount of food per consumer unit was converted into amounts of dietary energy (kcal) in two steps: (i) household measures were converted into grams (based on weightings done before and during the survey, complemented by conversion data from previous surveys in Kenya) and (ii) grams were converted into dietary energy value (kcal) using a food composition table (Platt 1962). The values of all the foods consumed during the recall period were added up to give the total dietary energy consumed per consumer unit during this period.

The dietary recall period covered the 48 hours immediately preceding the start of the interview, i.e. two days' consumption, also called consumption units (and not to be confused with consumer units). In cases where meals were recorded at household level before that period (e.g. breakfast or lunch of the day before yesterday), such meals were omitted from the calculations. When the recall period was shorter than two days due to missing data, the following deductions were made from the two consumption units: 0.17 for each missed breakfast, 0.33 for each missed lunch and 0.5 for each missed dinner. Thus, a household-specific denominator was obtained (usually two consumption units, but fewer in a number of cases due to missing data). The total dietary energy consumed per consumer unit during the recall period was divided by that denominator to obtain the estimated amount of dietary energy (kcal) per consumer unit per day in a particular household.

Table A6.5 Estimation of profitability of keeping dairy cows: The case of Baba Josephine

Capital costs	• animals 1 cow @ sh. 35,000	
	1 cow @ sh. 25,000	
	Total	sh. 60,000
	• shed[1]	sh. 10,000
	• milking utensils[2]	sh. 5,000
	Total	sh. 75,000
Variable costs (per year)	• grass:[3]	
	'rain season': 244 days * 5 bags = 1220 bags @ 25 sh. = sh. 30,500	
	'dry season': 122 days * 5 bags = 610 bags @ 50 sh. = sh. 30,500	
	Total	sh. 61,000
	• supplementary feeds:[4] sh. 7,270 per animal per year =	sh. 14,540
	• veterinary drugs:[5]	sh. 2,320
	• additional check-ups: 200 sh/animal =	sh. 400
	• milking salve: 100 sh/cow =	sh. 200
	• one herds boy @ sh. 1,500/month =	sh. 18,000
	Total	sh. 96,460
Income from milk sales[6] (per year)	Cow A: 'rain season': 210 days * 15 l/day = 3,150 litres	
	'dry season': 156 days * 7.5 l/day = 1,170 litres	
	Cow B: 'rain season': 210 days * 13 l/day = 2,730 litres	
	'dry season': 156 days * 6.5 l/day = 1,014 litres	
	Total milk production 8,064 litres	
	Sales: 'rain season': 2/3 * 8,064 l = 5,376 l @ 21 sh/l =	sh. 112,896
	'dry season': 1/3 * 8,064 l = 2,688 l @ 25 sh/l =	sh. 67,200
	Total income from milk sales	sh. 180,096

Notes:
1) Maximum costs for a fairly good and easily cleanable structure with a floor of cut stone or cement and a corrugated iron roof top, i.e. rain and sun proof.
2) Sieve (500 sh.), 10-litre stainless steel milking pail (2,500 sh.), 10-litre milk can (2,000 sh.).
3) Calculation based on:
   - a grass consumption of about 62.5 kg/day/animal
   - 1 bag = 25 kg of well-stuffed green grass, hence 62.5 kg = 2.5 bags
   - 'rainy season' = 8 months (244 days), 'dry season' = 4 months (122 days)
4) Dairy meal and mineral salts.
5) Includes dipping against ectoparasites (600 sh/animal); endoparasites (360 sh/animal); one round of vaccination by a private veterinary practioner (200 sh/animal).
6) Calculation based on:
   - lactation period of 210 days, i.e. full production; remaining 156 days half production
   - total milk production: 2/3 sold in 'rainy season', 1/3 sold in 'dry season'.

Sources: Versleijen 2002: 70-74 and technical information from William Keyah (personal communication, April 2004).

Annex 7: The support

Table A7.1 Livestock-keeping characteristics by type of assistance (%)

	(N=)	assistance from extension officer* (25)	assistance from neighbour/relative (11)	no assistance (85)
rearing system	within compound	52.0	36.4	38.8
	free range	32.0	45.5	55.3
	both	16.0	18.2	5.9
	Total	100	100	100
used inputs?	yes	100	100	84.7
	No	-	-	15.3
	Total	100	100	100
types of inputs: % yes	improved breeds	68.0	18.2	8.2
	veterinary drugs	96.0	45.5	36.5
	ethno-veterinary drugs	16.0	54.5	16.5
	feed supplements	92.0	81.8	47.1
	urban waste as feed	36.0	27.3	18.8
	crop residues as feed	68.0	63.6	48.2
	food left-overs	12.0	54.5	17.6
disposal of animals' waste**	use part or all of it for own crop cultivation	68.0	36.4	43.4
	give part or all of it to neighbours	28.0	27.3	8.4
	dump part or all of it in the street	20.0	27.3	37.3
	dump part or all of it in dustbins, pits, compound	4.0	27.3	13.3
	sell all of it	-.-	-.-	1.2

* Includes extension officers, programme officers and veterinary officers.
** Total exceeds 100% due to combined answers.
Source: 1999 survey.

Table A7.2 Problems with livestock keeping by type of assistance (%)

	(N=)	assistance from extension officer* (25)	assistance from neighbour/relative (11)	no assistance (85)
problems?	yes	96.0	100	85.9
	no	4.0	-	14.1
	Total	100	100	100
types of problems: % yes**	diseases	76.0	81.8	69.4
	theft	12.0	9.1	25.9
	lack of feed	36.0	9.1	9.4
	lack of funds/capital	12.0	9.1	10.6
	lack of safe drinking water	20.0	18.2	4.7
	predators	8.0	9.1	9.4
	lack of space	-.-	-.-	9.4
	harassment	12.0	9.1	3.5
	lack of market	8.0	9.1	1.2
	pests/parasites	4.0	9.1	1.2
	lack of labour	8.0	-.-	-.-
	nuisance	4.0	-.-	1.2

* Includes extension officers, programme officers and veterinary officers.
** Total exceeds 100% due to combined answers.
Source: 1999 survey.

Table A7.3 ARDP farmers and non-ARDP farmers: farming inputs (%)

		ARDP farmers (N=29)	non-ARDP farmers (N=48)
inputs for crop cultivation	manure	75.9	77.1
	crop residues	24.1	25.0
	chemical fertiliser	82.8	91.7
	chemical pesticides	60.7	66.7
	local seeds/seedlings	64.3	44.7
	improved seeds/seedlings	82.8	77.1
	irrigation	13.8	21.3
	extension services	55.2	20.8
inputs for livestock keeping	improved breeds/artificial insemination	48.3	35.0
	feed supplements	51.2	66.7
	veterinary drugs	82.2	68.8
	ethno-veterinary medicines	24.1	17.0
	extension services	37.9	39.6

Source: King'ori (forthcoming).

Table A7.4 ARDP farmers and non-ARDP farmers: major problems with farming (%)

		ARDP farmers (N=29)	non-ARDP farmers (N=48)
problems with crop cultivation	lack of water	24.1	45.8
	crop pests/diseases	13.7	25.0
	poor market	24.1	14.6
	theft	3.4	-.-
problems with livestock keeping	animal diseases	48.3	52.1
	lack of fodder	44.8	20.8
	high prices of inputs	31.0	20.8
	lack of safe drinking water	20.7	14.6
	lack of capital	3.4	25.0
	lack of space	17.2	8.3
	poor market	10.3	12.5
	lack of labour	10.3	6.3
	theft	6.9	6.3
	harassment	3.4	4.2

Source: King'ori (forthcoming).

Annex 8: The environment

Table A8.1 Heavy metal concentrations in the soils by sampling site (mg/l)

Sampling site↓ WHO standards→	Zn 50-150	Pb 5-10	Cd 0.05-0.2	Hg <1
1 Mwariki	29.8	2.5	0.19	0.09
2 Rhonda Sewage	241.0	4.5	0.21	0.10
3 London (dump site)	135.6	4.5	0.51	0.61
4 Lanet Free Area	3.0	1.0	0.09	0.11
5 Rhonda Pondamali	10.0	3.2	0.01	0.07
6 Lanet	6.7	1.0	0.01	0.01
7 Kivumbini	21.1	3.1	0.12	0.30
8 Rhonda Kaptembwa	15.0	2.1	0.09	0.02
9 Kabachia	26.2	1.9	0.08	0.13
10 Rhonda Muslim	31.0	1.1	0.11	0.11
11 Kaloleni	53.5	1.4	0.08	0.14
12 Milimani	3.1	1.8	0.08	0.09

Source: Nyandwaro (forthcoming).

Table A8.2 Heavy metal concentrations in Amaranthus (spinach) by sampling site (mg/l)

Sampling site↓ WHO standards→	Zn 50-150	Pb 5-10	Cd 0.05-0.2	Hg <1	Observed plant characteristics
1 Mwariki	26.8	10.3	0.02	0.10	Leafy, mature plants
2 Rhonda Sewage	210.0	10.0	0.10	0.21	Very green, leafy, young plants
3 London (dump site)	130.0	4.6	0.09	0.50	Mixture of very green plants and withered ones
4 Lanet Free Area	2.5	2.3	0.08	0.09	Healthy plants
5 Rhonda Pondamali	8.0	2.0	0.07	0.08	Fully grown plants with tiny leaves
6 Lanet	4.6	1.8	0.07	0.01	Healthy plants
7 Kivumbini	17.1	1.8	0.08	0.11	Very leafy, soft, young plants
8 Rhonda Kaptembwa	9.5	2.0	0.04	0.08	Very green, mature plants with soft stem
9 Kabachia	4.7	7.8	0.11	0.09	Healthy, mature plants
10 Rhonda Muslim	2.2	2.3	0.12	0.11	Very green plants with soft leaves
11 Kaloleni	3.2	2.5	0.07	0.10	Very green, soft, young plants
12 Milimani	4.1	1.8	0.06	0.05	Healthy young plants

Source: Nyandwaro (forthcoming).

Annex 9: The schools

Table A9.1 School characteristics

	Name of school	Area income level	School category	School type	Farming?	Compound size (acres)	Plot size (acres)	Lunch provided?
1	Lanet Secondary	pu	s	d	c	20	6.5	1
2	Lanet Primary	pu	p	d	c	8	4	-
3	Nakuru East	pu	p	d	c	11.5	3	1
4	Rhino	pu	p	d	-	-	-	-
5	St. George's Muthaiti	m	p	d/b	c/l	33	25	1
6	Ruth Wangugi	pu	p	d	-	-	-	1
7	Nairobi Road	pu	p	d	-	-	-	-
8	Shinners Academy	m	s	b	c	5	1	1
9	Madaraka	m	p	d	c	7	1.5	-
10	Jamuhuri	m-h	p	d	-	-	-	1
11	Kisulisuli	m	p	d	-	-	-	-
12	Bondeni	l-m	p	d	c/l	12	1	1
13	St. Columbus	m-h	s	d/b	c	12	2	1
14	St. Michaels	m	s	b	-	-	-	1
15	Nakuru High	h	s	d/b	l	-	-	1
16	Upper Hill	m-h	s	d	c/l	10	2	1
17	Loreto Nakuru	m-h	s	d	c	6	2.5	1
18	Rasul al Akram Acad.	pu	p/s	b	c/l	?	8	1
19	Lion Hill	m	p	d	c	10	0.5	-
20	Hyrax	m-h	p	d	c	6	1	-
21	Nakuru Blankets	m-h	s	d/b	c	15	3	1
22	Lake Nakuru High	m-h	s	d/d	c/l	5	1.5	1
23	Kenyatta	m	p	d	-	-	-	-
24	Khalsa High	m	s	d	-	-	-	-
25	St. Joseph	m	p	d	c	0.9	0.3	1
26	Christ the King Acad.	m	s	b	c/l	10	2	1
27	Heshima	l	p	d	c	5	0.5	1
28	Kaptembwa	l	p	d	c	3	0.5	-
29	Nakuru West	l-m	p	d	c	3.8	0.5	1
30	Koinange	l-m	p	d	c	2.5	0.125	1
31	St. Nicholas Faith Acad.	l-m	p	d/b	c/l	3	0.25	1
32	Kimathi	l-m	p	d	-	-	-	-
33	Lake View	l-m	p	d	-	-	-	-
34	Pangani	p	d	8	c	2	-	-
35	Langalanga	l-m	p	d	-	-	-	-
36	Freehold	l-m	p	d	c	5	0.125	1
37	Umoja	l-m	p	d	-	-	-	-
38	St. Xavier's	m	p	d	c	?	0.04	1
39	St. Xavier's	m	s	d	c	?	0.08	1
40	Menengai	m-h	p	d	l	-	-	1
41	Afraha	m	s	d	c	?	0.05	1
42	Kaloleni	l-m	p	d	-	-	-	-
43	St. Teresa	l	p	d	-	-	-	-
44	Nakuru Primary	m	p	d	c	11	4	-
45	Baharini	l-m	p	d	-	-	-	-
46	Flamingo	l-m	s	d	c	1.5	0.5	1
47	Ngala Special	m	p	d/b	c/l	7	3.5	1

>>>

Table A9.1, continued

	Name of school	Area income level	School cate-gory	School type	Farm-ing?	Com-pound size (acres)	Plot size (acres)	Lunch pro-vided?
48	St. Mary's	m-h	p	d	c	7	2	1
49	Langalanga	m	s	d	c	6	1	1
50	Menengai	m-h	s	d	c/l	13	2.25	1
51	Race Track	m	p	d	-	-	-	-
52	Mwariki	l	p	d	-	-	-	-
53	Khalsa	m	p	d	c	3	0.5	1
54	Kariba Road	m	p	d	-	-	-	1
55	Jull Jeady Girls	m	s	d/b	c	2.5	0.75	1
56	Eileen Ngochoch	l	p	d	-	-	-	-
57	Kiptenden	l	p	d	-	-	-	1
58	Baruti	pu	p	d	c	8	0.5	-
59	Archbishop Ndingi Acad.	l	p	d/b	-	-	-	1
60	Rhonda	l	p	d	c	8	1.5	-
61	Moi Primary	h	p	d	c	8	1	1
62	Moi Secondary	h	s	d	c/l	?	2	1
63	Prisons	p	d	10	c	4	-	
64	Hill Special	h	p	b	c/l	5	0.25	1
65	Milimani	h	p	d	-	-	-	-
66	Kenyatta	m	s	d	c	2.5	0.5	1
67	Nakuru Day	m	s	d	c	20	0.4	1
68	Uhuru	pu	p	d	c	10	1	1
69	Crater	m-h	p	d	-	-	-	-
70	St. John's	m-h	p	d	-	-	-	-
71	Victonell	h	p	d	-	-	-	1
72	Lenana	h	p	d	-	-	-	-
73	Malvin Jones	m-h	p/s	d	-	-	-	-
74	Flamingo	l-m	p	d	-	-	-	-
75	Nakuru West	l	s	d	c	8	0.75	1
76	Muslim	l-m	p	d	c	3	?	-
77	Park View	pu	p	d	-	-	-	1
78	Kigonor	pu	p	d	c	?	3	1
79	Kelelwet	pu	p	d	-	-	-	1
80	Mama Ngina	m	p	d	c	8	0.25	1

KEY:

Area income level
l = low
l-m = low to medium
m = medium
m-h = medium to high
h = high
pu = peri-urban

School category
p = primary
s = secondary
p/s = both primary and secondary

School type
d = day school
b = boarding school
d/b = both day and boarding

Farming?
c = crop cultivation
l = livestock keeping

Lunch provided?
1 = yes

Source: Odera (forthcoming).

Annex 10: The poor

Table A10.1 All households: household size by income class (%)

	up to Ksh. 5,000 (N=310)	more than Ksh. 10,000 (N=106)
1 member	17.4	1.9
2-4 members	59.7	34.9
5-7 members	18.4	49.1
8 or more members	4.5	14.2
Total	100	100

Source: 1999 survey.

Table A10.2 Non-farming households: reasons for *not* farming in town by type of farming and income class (%)

		crop cultivation		livestock keeping	
		<=5,000 (N=267)	>10,000 (N=50)	<=5,000 (N=267)	>10,000 (N=50)
land issues	no access to urban land	86.5	82.0	76.8	72.5
	have access to rural land	5.6	16.0	4.1	10.1
lack of other resources	no capital	33.7	14.0	32.2	7.2
	lack of time	6.4	10.0	5.2	10.1
	no labour available	1.9	-.-	2.6	7.2
legal consider-ations	harassment	2.2	2.0	3.0	11.6
	not allowed in town	-.-	4.0	-.-	1.4
	disapprove of it	2.6	12.0	3.4	15.9
	landlord disapproves	1.9	-.-	4.1	7.2
	neighbours complain	-.-	-.-	0.4	1.4
	theft/insecurity	1.9	-.-	2.6	1.4
other reasons	not worthwhile	3.7	10.0	6.0	7.2
	had not thought about it	0.4	2.0	1.5	2.9
	destruction by animals	0.7	-.-	0.7	1.4
	diseases	-.-	-.-	2.2	1.4
	was not in Nakuru	2.6	-.-	2.2	-.-

Note: Total exceeds 100%.
Source: 1999 survey.

Table A10.3 Rural-farming households: rural-farming characteristics by income class (%; plots)

		up to Ksh. 5,000 (N=216)	more than Ksh. 10,000 (N=98)
Plot size* (acres)	less than 1	20.8	12.2
	1 - 1.9	28.7	19.4
	2 - 4.9	31.0	30.6
	5 – 9.9	13.9	19.4
	10+	5.6	18.4
	Total	100	100
Who uses plot?	myself	49.1	71.0
	other family	38.4	19.0
	somebody else	8.1	5.0
	nobody	4.5	5.0
	Total	100	100
Importance of rural plot	food source	37.2	30.6
	income source	7.2	8.2
	food and income source	35.9	43.9
	no food and/or income source	19.7	17.3
	Total	100	100

* Average plot sizes are 3.0 and 6.2 acres, respectively.
Source: 1999 survey.

Table A10.4 Urban crop cultivators: plot characteristics by income class (%; plots)

		up to Ksh. 5,000 (N=46)	more than Ksh. 10,000 (N=61)
location	riverside	-.-	1.6
	roadside	8.7	3.3
	railway side	13.0	-.-
	under power line	4.3	-.-
	in own compound	45.7	68.9
	within estate	17.4	21.3
	in other estate	10.9	3.3
	elsewhere	-.-	1.6
	Total	100	100
distance to plot (minutes walking)	less than 10 minutes	58.0	83.6
	10-30 minutes	26.0	8.2
	30-60 minutes	10.0	6.6
	more than 60 minutes	6.0	1.6
	Total	100	100
plot size in m^2	<50	38.0	26.2
	50-99	6.0	13.1
	100-499	14.0	21.3
	500-999	12.0	8.2
	1000+	30.0	31.1
	Total	100	100
ownership of plot	own land	20.4	47.5
	landlord	49.0	41.0
	government	16.3	6.6
	other	10.2	4.9
	don't know	4.1	-.-
	Total	100	100

Source: 1999 survey.

Table A10.5 Urban crop cultivators: characteristics of crop cultivation by income class

		up to Ksh. 5,000 (N=43)	more than Ksh. 10,000 (N=56)
% households cultivating …	maize	79.1	58.9
	beans	74.4	50.0
	kales	58.1	75.0
	spinach	20.9	28.6
	onions	25.6	28.6
	tomatoes	11.6	28.6
	cowpeas	16.3	19.6
	bananas	7.0	23.2
	irish potatoes	23.3	21.4
	saget	9.3	16.1
person responsible for crop cultivation	household head	51.2	19.6
	spouse	41.9	53.6
	other household member	4.6	14.3
	hired labour	-.-	3.6
	other	2.3	1.8
	Total	100	100
crop cultivation full-time occupation? (% yes)		25.6	10.7
used hired labour for crop cultivation? (% yes)		18.6	35.7
% households using …	chemical fertiliser	41.9	41.1
	manure as fertiliser	41.9	66.1
	crop residue as fertiliser	20.9	41.1
	urban waste as fertiliser	-.-	3.6
	chemical pesticides	32.6	35.7
	chemical insecticides	11.6	10.7
	local seeds/seedlings	60.5	71.4
	improved seeds/seedlings	53.5	64.3
	irrigation	23.3	57.1

Source: 1999 survey.

Table A10.6 Urban crop cultivators: problems with crop cultivation by income class (%)

	up to Ksh. 5,000 (N=43)	more than Ksh. 10,000 (N=56)
no problem	18.6	16.1
theft of crops	44.2	33.9
inadequate rainfall	30.2	26.8
destruction by animals	23.3	17.9
pests/insects	23.3	23.2
lack of water for irrigation	4.7	17.9
diseases	7.0	12.5
lack of inputs/capital	11.6	8.9
harassment	2.3	-.-
bad quality seeds	-.-	1.8
poor soil	2.3	1.8
lack of space/land	-.-	3.6
lack of labour	2.3	-.-
weeds	-.-	1.8
too much rainfall	-.-	1.8
poor seasonal timing	-.-	1.8
sewage burst	-.-	1.8

Note: Totals exceed 100%. Source: 1999 survey.

Table A10.7 Urban livestock keepers: characteristics of livestock keeping by income class (%)

		up to Ksh. 5,000 (N=43)	more than Ksh. 10,000 (N=37)
% households keeping* …	cattle	11.6	37.8
	sheep	4.7	10.8
	goats	7.0	21.6
	pigs	-.-	-.-
	chicken	90.7	78.4
	ducks	9.3	-.-
	rabbits	2.3	-.-
	doves	2.3	-.-
	turkeys	-.-	8.1
rearing system	within compound	35.9	48.3
	free range	64.1	44.8
	both	-.-	6.9
	Total	100	100
person responsible for livestock keeping	household head	48.8	27.0
	spouse	48.8	62.2
	other household member	-.-	5.4
	hired labour/herds boy	2.3	5.4
	Total	100	100
livestock keeping full-time occupation? (% yes)		18.6	16.2
used hired labour for livestock keeping? (% yes)		14.0	43.2
% households using …	improved breeds	9.3	40.5
	veterinary drugs	23.3	81.1
	feed supplements	39.5	86.5
	urban waste as feed	20.9	24.3
	crop residues as feed	53.5	62.2
	ethno-veterinary drugs	23.3	13.5
	food left-overs	23.3	13.5
disposal of the animals' waste*	use part or all for own crop cultivation	25.6	70.3
	give part or all to neighbours	9.3	16.2
	dump part or all in the street	48.8	16.2
	dump part or all in compound/pit	16.3	5.4
	sell all of it	-.-	2.7
	not accumulated	4.7	-.-

* Total exceeds 100%.
Source: 1999 survey.

Table A10.8 Urban livestock keepers: problems with livestock keeping by income class (%)

	up to Ksh. 5,000 (N=43)	more than Ksh. 10,000 (N=37)
no problem	11.6	5.4
diseases	69.8	83.8
theft	18.6	24.3
lack of feed	11.6	18.9
lack of funds/capital	9.3	10.8
lack of safe drinking water	9.3	16.2
predators	7.9	10.8
lack of space	14.0	-.-
harassment	9.3	8.1
lack of market	2.3	8.1
pests/parasites	-.-	2.7
lack of labour	-.-	2.7
nuisance	-.-	2.7

Note: Totals exceed 100%.
Source: 1999 survey.

Table A10.9 Urban farming households: general food security issues by type of farming and income class (%)

		crop cultivation		livestock keeping	
		=<Ksh. 5,000 (N=43)	>Ksh. 10,000 (N=56)	=<Ksh. 5,000 (N=43)	>Ksh. 10,000 (N=37)
Reasons to farm in town*	needed food	95.3	100	90.7	86.5
	needed income	16.3	10.7	30.2	27.0
	to diversify income	16.3	5.4	20.9	35.1
	hobby/custom	4.7	12.5	7.0	10.8
	other reasons	4.7	-.-	-.-	-.-
Importance of urban farming*	could not survive without it	14.0	7.1	2.3	2.7
	major food source	20.9	16.1	16.3	16.2
	major income source	4.7	-.-	11.6	8.1
	additional food source	55.8	73.2	53.5	48.7
	additional income source	9.3	5.4	23.3	27.0
	could do without it	2.3	3.6	4.7	5.4

* Total exceeds 100%.
Source: 1999 survey.

Table A10.10 Urban farming households: general food security issues by year and income class (%)

		1998		1999	
		=<Ksh. 5,000	>Ksh. 10,000	=<Ksh. 5,000	>Ksh. 10,000
		(N=68)	(N=63)	(N=13)	(N=29)
Always enough to eat?	yes always	88.2	96.8	23.1	75.9
	most of the time	7.4	1.6	46.2	10.3
	about half of the time	1.5	-.-	23.1	13.8
	now and then	2.9	-.-	7.7	-.-
	never	-.-	1.6	-.-	-.-
	Total	100	100	100	100
Most important food source in 1998	own urban production	5.9	6.3	-.-	3.2
	urban + rural production	8.8	17.5	-.-	-.-
	urban production + purchased	26.5	25.4	15.4	32.3
	urban production, rural production + purchased	-.-	1.6	-.-	-.-
	purchased	42.6	33.3	84.6	48.4
	rural production	-.-	1.6	-.-	-.-
	rural production + purchased	16.2	14.3	-.-	16.1
	Total	100	100	100	100
		(N=43)	(N=56)	(N=10)	(N=26)
Contribution of urban crop cultivation to household food consumption	most of the food	20.9	14.3	10.0	7.7
	about half of the food	39.5	17.9	10.0	15.4
	less than half of the food	14.0	23.2	20.0	23.1
	small portion	25.6	32.1	40.0	46.2
	negligible	-.-	10.7	-.-	7.7
	none	-.-	1.8	20.0	-.-
	Total	100	100	100	100

Source: 1999 survey.

Table A10.11 Characteristics of urban plots by sex of household head (%)

			male	female
		(N=; plots)	(153)	(26)
size of plot (m^2)	<100		38.6	46.2
	100-999		26.8	38.4
	1000+		34.6	15.3
	Total		100	100
location of plot	in own compound		59.7	65.4
	within residential estate		15.4	23.1
	elsewhere		24.9	11.5
	Total		100	100
ownership of plot	own land		32.7	34.6
	landlord		45.1	50.0
	government land		15.7	3.8
	other		6.5	11.6
	Total		100	100
distance to plot	<10 minutes walking		67.5	88.5

Source: 1999 survey.

Annex 11: Results of the NUAP workshop, 27-28 November 2002

In November 2002, a two-day workshop was held in Nakuru town. The aim was threefold: (1) to present the results of the various NUAP studies to the local stakeholders, (2) to discuss the practices of urban agriculture in Nakuru Municipality and formulate possible improvements, and (3) to discuss policy and planning issues concerning urban agriculture in Nakuru Municipality and formulate recommendations. This annex contains a summary of the latter two objectives.[1]

Possible improvements of urban agriculture in Nakuru Municipality

Type of farming	Practices/-methods	Problems	Possible improvements
crop cultivation	road reserves	- insecurity - obstruction - contamination	- train farmers on alternative methods of farming
	undeveloped plots, open spaces	- losses	- encourage sausage farming
	developed plots	- parasites and diseases	- multi-storey farming - short plants requiring minimum distances
	river banks	- soil erosion - chemical pollution - diversion of river - loss of biodiversity	- educate farmers
	near/along sewer works	- costly - public health concerns	- rehabilitation of the sewage system - more research on health hazards - need for more sensitisation

Continues on next page

[1] Annex 11 is based on the workshop report compiled by Samson W. Mwangi who is gratefully acknowledged for that.

Type of farming	Practices/-methods	Problems	Possible improvements
large livestock	free range	- safety hazards - nuisance - pollution - destructive - health problems - MCN by-laws	- zoning - ban in built-up areas - restrict to own land
	zero-grazing	- cost of feed - waste disposal - hygiene - labour	- link with crop cultivation - mobilisation for composting
poultry	free range	- losses, e.g. theft predators, diseases, poor production	- extension services - enclose - alternative low-cost feeds - sensitisation
	enclosed	- waste disposal - initial capital - feeding, low labour - nuisance, noise - easy to steal - hygiene - marketing	- link with crop cultivators - cooperative farming - explore cheap enclosure costs - proper timing
small livestock	free range or enclosed	- sharing room with the family - health concerns	- educate farmers - educate on the usefulness of rearing rabbits (little space needed)

Policy and planning recommendations

The discussion concerning policy and planning recommendations was guided by four "entry points for planning and policy issues in urban agriculture", as formulated by Shingirayi Mushamba[2] in his keynote address to the workshop (Mushamba 2002: 3-5):

1) Official policy and planning controls exist for urban agriculture, but such controls are increasingly a minor part of the full urban agriculture 'industry'.
2) A certain paralysis can be observed in the progress to integrate urban agriculture, especially at the planning level. (…) Local officials have heard, have seen urban agriculture, and most officials either do it or see it every day, but not much has been done to integrate urban agriculture into current plans (such as Localising Agenda 21).
3) Changes in policy have far lagged behind practice on the ground and policy changes have not been followed by changes in planning practice. (…) Urban agriculture has been in existence in spite of denials by policy makers and planners. (…) Although there have

2 Senior Programme Officer and Urban Agriculture Programme Coordinator, Municipal Development Partnership (MDP), Harare.

been some policy shifts (…) these were not followed by the necessary planning changes (as was the case in other small-scale industrial sectors).

4) In most countries in the East African region, there exists no comprehensive policy specifically geared towards urban agriculture. (…) There is no coherent view among the different municipal departments on urban agriculture, its role, its negative and positive impacts and about how it should be regulated, which makes it more complicated to reach consensus.

During the discussions, it was observed that the current policies are outdated with respect to the dynamics of urban agriculture. There is a discrepancy between policy and practice and there is no clarity in terms of actual policy regarding farming in town:

- The Local Government Act and the Agriculture Act seem to contradict each other in issues related to urban agriculture.
- Operational definitions as contained in various legislations do not render sufficient weight to urban agriculture. The preference is on other types of urban development.
- The Physical Planning Act deals with development and control of physical infrastructure and does not talk about agriculture at all. There is no provision for urban agriculture in the Nakuru Strategic Structure Plan, a blueprint to guide development in Nakuru until the year 2025. In spite of that, there is plenty of opportunity through area-based action plans to incorporate urban agriculture.
- The presence of urban agriculture within Nakuru Municipality is acknowledged, but there is a lack of policy to support it. Urban agriculture will continue to grow but will change its nature determined by the growth of the built environment.
- There is likely to be competition between different types of land use in the town and there might be a reduction in urban agriculture in the future as other physical developments will continue. On-plot agriculture is likely to replace peri-urban farming. There will be a tendency to move towards different types of urban agriculture, especially where there is limited space, e.g. roof-top farming, use of containers and use of space along the highways.
- There is need for a mechanism to control and monitor urban agriculture projects in all human settlements.

The following recommendations were formulated:

- There is need to control urban agriculture and make the situation better for those who rely on urban farming as a means of livelihood. Therefore, policies are required that will link urban agriculture with other means of livelihood. There is need to create zones in some places that could be utilised for urban agriculture, especially in the non-productive areas. Proper zoning, regulations and guidelines need to be put in place, especially on what to produce and where.
- There is also need to control and manage adverse effects of urban farming. Appropriate legal guidelines should be put in place and a very clear policy in urban agriculture should be formulated by all stakeholders. There should be proper consultations of all actors and there is need for awareness raising.

• The Ministry of Agriculture and the Municipal Council should educate people involved in urban agriculture and there is need for awareness raising on space optimisation and appropriate production methods.

• There is need for guidelines for waste management that is generated by urban farming activities. There is also need for waste management strategies by the local government.

• Local authorities and other policy-making institutions need to look at urban agriculture as an important economic activity. Research and training institutions need to raise awareness on both the benefits and the adverse effects of farming in town.

• There is need to have a multi-stakeholder task force in the Municipal Council of Nakuru (MCN) to deal with issues related to urban agriculture. Urban farmers should also be included in this task force.

• Zonal Development Committees need to discuss urban farming activities within their respective zones and develop strategies how to undertake such activities.

• A follow-up meeting of the workshop should be made by the NUAP team to the MCN and other stakeholders and this should be done when the new Council is constituted.

References

AJAEGBU, H.I., D. GROSSMAN & L.M. VAN DEN BERG (2000), *Market gardening, urban growth and sustainable income generation on the Jos Plateau, Nigeria*. Amsterdam: Royal Tropical Institute, NIRP Research for Policy Series No.3.

AMEND, J. & E. MWAISANGO (1998), *Status of soil contamination and soil fertility – The case of urban agriculture in Dar es Salaam*. Dar es Salaam: Urban Vegetable Promotion Project.

ARDP (1998), *Quarterly Report 2*. Nakuru: Catholic Diocese of Nakuru, Agriculture and Rural Development Programme.

ARMAR-KLEMESU, M. (2000), Urban agriculture and food security, nutrition and health. In: N. Bakker *et al*., eds., *Growing cities, growing food. Urban agriculture on the policy agenda*, pp. 99-118. Feldafing (Germany): Deutsche Stiftung für internationale Entwicklung (DSE).

ARMAR-KLEMESU, M. & D. MAXWELL (2000), Accra: Urban agriculture as an asset strategy, supplementing income and diets. In: N. Bakker *et al*., eds., *Growing cities, growing food. Urban agriculture on the policy agenda*, pp. 183-208. Feldafing (Germany): Deutsche Stiftung für internationale Entwicklung (DSE).

ATAKUNDA, G. & D. MAXWELL (1996), Farming in the City of Kampala: Issues in urban management. *African Urban Quarterly* 11(2-3): 261-272.

ATEKA, E. (1999), Farm Management Officer, Ministry of Agriculture, Nairobi, personal communication.

BAKKER, N., M. DUBBELING, S. GUENDEL, U. SABEL-KOSCHELLA & H. DE ZEEUW, eds. (2000), *Growing cities, growing food. Urban agriculture on the policy agenda*. Feldafing (Germany): Deutsche Stiftung für internationale Entwicklung (DSE).

BAXTER, J. (1994), *Urban agriculture in South Africa: General perspectives and local experience*. Johannesburg: University of Witwatersrand.

BEALL, J. (2002), Living in the present, investing in the future – Household security among the urban poor. In C. Rakodi, with T. Lloyd-Jones, eds, *Urban livelihoods: A people-centred approach to reducing poverty*, pp. 71-87. London: Earthscan Publications.

BEBBINGTON, A. (1999), Capitals and capabilities: A framework for analyzing peasant viability, rural livelihoods and poverty. *World Development* 17(12): 2021-2044.

BIGSTEN, A. & S. KAYIZZI-MUGERWA (1992), Adaption and distress in the urban economy: A study of Kampala households. *World Development* 20(10): 1423-1441.

BLAIKIE, P., T. CANNON, I. DAVIS & B. WISNER (1994), *At risk. Natural hazards, people's vulnerability and disasters*. London: Routledge.

BOWYER-BOWER, T.A.S. (2002), Environmental consequences of illegal urban agriculture in African cities. Paper presented at the Fifth International Eco-City Conference, Shenzhen, China, August 19-23, 2002.

BOWYER-BOWER, T.A.S. & D. DRAKAKIS-SMITH (1996), *The needs of the urban poor versus environmental conservation: Conflict in urban agriculture*. Final Report of ODA/ESCOR Project R5946. London: Overseas Development Administration.

BROCK, B. (1999), Actual and potential contribution of urban agriculture to environmental sanitation: A case study in Cotonou. In O. Smith, ed., *Urban agriculture in West Africa: Contributing to food security and urban sanitation*, pp. 126-137. Ottawa: International Development Research Centre (IDRC).

BROCK, B. & D. FOEKEN, D. (2005), Urban horticulture for a better environment. A case-study of Cotonou, Benin. *Habitat International* (yet to be published).

BROWN, A. & T. LLOYD-JONES (2002), Spatial planning, access and infrastructure. In C. Rakodi, with T. Lloyd-Jones, eds, *Urban livelihoods: A people-centred approach to reducing poverty*, pp. 188-204. London: Earthscan Publications.

BURGESS, R., M. CARMONA & T. KOLSTEE (1996), *Contemporary urban strategies and urban design in developing countries*. The Hague: Ministry of Foreign Affairs, Directorate General International Cooperation.

BYERLEY, A. (1996), *Urban agriculture in Botswana: A preliminary investigation of extent, issues and potential (a minor field study)*. Uppsala: Swedish University of Agricultral Sciences.

CARNEY, D. (1998), Implementing the sustainable rural livelihood approach. In D. Carney, ed., *Sustainable rural livelihoods: What contribution can we make?*, pp. 3-23. London: Department for International Development (DFID).

CARNEY, D. (1999), *Social capital. Key sheets for sustainable livelihoods. Policy planning and implementation*. London: DFID/ODI (http://www.oneworld.org/odi/keysheets)

CHAMBERS, R. (1983), *Rural development: Putting the last first*. Harlow: Longman.

CHAMBERS, R. (1995), *Poverty and livelihoods: Whose reality counts?* Brighton: University of Sussex, Institute of Development Studies.

CHAMBERS, R. & G. CONWAY (1992), *Sustainable rural livelihoods: Practical concepts for the 21st century*. Brighton: University of Sussex, Institute of Development Studies, Discussion Paper 296.

CORNISH, G. (2002a), Informal irrigated production in peri-urban Nairobi and Kumasi. Paper presented at the Regional Workshop on "Urban Policy Implications of Enhancing Food Security in African Cities", Nairobi, 27-31 May, 2002.

CORNISH, G. (2002b), contribution to e-conference on the use of untreated urban waste water in agriculture, June 24, 2002.

DEELSTRA, T. & H. GIRARDET (2000), Urban agriculture and sustainable cities. In: N. Bakker *et al.*, eds., *Growing cities, growing food. Urban agriculture on the policy agenda*, pp. 43-66. Feldafing (Germany): Deutsche Stiftung für internationale Entwicklung (DSE).

DE HAAN, L.J. (2000), *Livelihood, locality and globalisation*. Nijmegen: Nijmegen University Press.

DE HAAN, L. & A. ZOOMERS (2003), Development geography at the crossroads of livelihood and globalisation. *Tijdschrift voor Economische en Sociale Geografie* 94(3), 350-362.

DE MEULDER, B. (1998), Nakuru, a patchwork. From colonial outpost to East African town. In: B. de Meulder et al., *Nakuru. An African town*, pp. 13-34. Leuven: City of Leuven/Catholic University.

DENNERY, P. (1995), *Inside urban agriculture: an exploration of food producer decision making in a Nairobi low income area*. Wageningen: Agricultural University (MSc thesis).

DENNERY, P. (1996), Urban producers' decision-making in Kibera, Nairobi, Kenya. *African Urban Quarterly* 11(2-3): 189-200.

DEVAS, N. (2002), Urban livelihoods – Issues for urban governance and management. In: C. Rakodi, with T. Lloyd-Jones, eds., *Urban livelihoods: A people-centred approach to reducing poverty*, pp. 205-221. London: Earthscan Publications.

DE ZEEUW, H., S. GUENDEL & H. WAIBEL (2000), The integration of agriculture in urban policies. In: N. Bakker *et al.*, eds., *Growing cities, growing food. Urban agriculture on the policy agenda*, pp. 161-180. Feldafing (Germany): Deutsche Stiftung für internationale Entwicklung (DSE).

DIALLO, S. (1993), *Urban agriculture research in West Africa: Record, capacities and opportunities*. Ottawa: IDRC, Cities Feeding People Series, Report no. 5.

DONGUS, S. (2000), *Vegetable production on open spaces in Dar es Salaam – Spatial changes from 1992 to 1999*. Dar es Salaam: Urban Vegetable Promotion Project.

DRAKAKIS-SMITH, D. (1992), Strategies for meeting basic food needs in Harare. In J. Baker and P. O. Pedersen, eds., *The Rural-Urban Interface in Africa: Expansion and Adaptation*, pp. 258-283. Uppsala: Nordiska Afrikainstitutet.

DRAKAKIS-SMITH, D., T.A.S. BOWYER-BOWER & D. TEVERA (1995), Urban poverty and urban agriculture: An overview of the linkages in Harare. *Habitat International* 19(2): 183-194.

DRESCHER, A.W. (1996), Urban microfarming in central Southern Africa: A case study of Lusaka, Zambia. *African Urban Quarterly* 11(2-3): 226-245.

DUCHHART, I. & F. GROOTENHUIS (1993), Urban agriculture, incentive planning and Green Towns. *Urban Perspectives*, 4(1): 9-10.

EBERHARD, R. (1989), *Urban agriculture: The potential in Cape Town: Literature review*. City of Cape Town: Town Planning Branch, Working Paper 89/E2.

EGAL, F., A. VALSTAR & S. MEERSHOEK (2001), Urban agriculture, household food security and nutrition in Southern Africa. Paper presented at the Sub-Regional Expert Consultation "Urban and Peri-Urban Horticulture in Southern African Countries", Stellenbosch, South Africa, 16-18 January 2001.

EGZIABHER, A.G. (1994), Urban farming, cooperatives, and the urban poor in Addis Ababa. In A.G. Egziabher *et al.*, *Cities feeding people: An examination of urban agriculture in East Africa*, pp. 85-104. Nairobi/Ottawa: IDRC.

EGZIABHER, A.G., D. LEE-SMITH, D.G. MAXWELL, P.A. MEMON, L.J. MOUGEOT & C.J. SAWIO (1994), *Cities feeding people. An examination of urban agriculture in East Africa*. Ottawa: International Development Research Centre.

ELLIS, F. (2000), *Rural livelihoods and diversity in developing countries*. Oxford: Oxford University Press.

ENDA-ZIMBABWE (1996), *Urban agriculture in Harare: Results and recommendations of a household survey conducted in Harare*. Harare: ENDA-Zimbabwe.

FOEKEN, D. (2005), *Urban agriculture in East Africa as a tool for poverty reduction: A legal and policy dilemma?* Leiden: African Studies Centre, ASC Working Paper 65.

FOEKEN, D. & A.M. MWANGI (1998), Does access to land have a positive impact on the food situation of the urban poor? A case-study in Nairobi. *East African Social Science Research Review*, 14(1): 19-32.

FOEKEN, D. & A.M. MWANGI (2000), Increasing food security through urban farming in Nairobi. In: N. Bakker *et al.*, eds., *Growing cities, growing food. Urban agriculture on the policy agenda*, pp. 303-328. Feldafing (Germany): Deutsche Stiftung für internationale Entwicklung (DSE).

FOEKEN, D. & S.O. OWUOR (2000a), *Urban farmers in Nakuru, Kenya*. Leiden: African Studies Centre, ASC Working Paper no. 45 (also on www.cityfarmer.org/nakuru.html)

FOEKEN, D. & S.O. OWUOR (2000b), Livestock in a middle-sized East-African town: Nakuru. *Urban Agriculture Magazine* 1(2): 20-22.

FOEKEN, D. & S.O. OWUOR (2001), Multi-spatial livelihoods in Sub-Saharan Africa: Rural farming by urban households — The case of Nakuru town, Kenya. In M. de Bruijn, R. van Dijk & D. Foeken, eds., *Mobile Africa: Changing patterns of movement in Africa and beyond*, pp. 125-140. Leiden: Brill.

FOEKEN, D., S.O. OWUOR & W. KLAVER (2002), *Crop cultivation in Nakuru town, Kenya: Practice and potential*. Leiden: African Studies Centre, ASC Working Paper no. 50 (also on http://www.ascleiden.nl/pdf/workingpaper50.pdf).

FOEKEN, D., M. SOFER & M. MLOZI (2004), *Urban agriculture in Tanzania: Issues of sustainability*. Leiden: African Studies Centre, Research Report 75.

FREEMAN, D.B. (1991), *A city of farmers: Informal urban agriculture in the open spaces of Nairobi, Kenya*. Toronto: McGill University Press.

FREEMAN, D.B. (1993), Survival strategy or business training ground? The significance of urban agriculture for the advancement of women in African cities. *African Studies Review*, 36(3): 1-22.

GATHURU, P.K. (1993), Women in sustainable urban improvement: Urban farming in an informal settlement of Nairobi. *Urban Perspectives* 4(1): 14-24.

GBADEGESIN, A.S. (1991), Farming in the urban environment of a developing nation. A case study from Ibadan metropolis in Nigeria. *The Environmentalist* 11: 105-111.

GEFU, J.O. (1992), Part-time farming as an urban survival strategy: A Nigerian case study. In: J. Baker & P.O. Pedersen, eds., *The rural-urban interface in Africa: Expansion and adaptation*, pp. 295-302. Uppsala: The Scandinavian Institute of African Studies.

GONZALEZ NOVO, M. & C. MURPHY (2000), Urban agriculture in the City of Havana: A popular response to a crisis. In: N. Bakker *et al.*, eds., *Growing cities, growing food. Urban agriculture on the policy agenda*, pp. 329-347. Feldafing (Germany): Deutsche Stiftung für internationale Entwicklung (DSE).

GUMBO, D. & T.W. NDIRIPO (1996), Open space cultivation in Zimbabwe: A case study of Greater Harare. *African Urban Quarterly* 11(2-3): 210-214.

HOON, P., N. SINGH & S. WANMALI (1997), *Sustainable livelihoods: Concepts, principles and approaches to indicator development*. New York: UNDP, Bureau for Development Policy, Social Development and Poverty Eradication Division, Discussion Paper (http://www.undp.org/sl/publication/ind2.htm)

ISIKA, M., F.M. RUGENYI, E. MUTISYA & M. GATHUMBI (2002), Review of current situation on urban and peri-urban aghriculture in Nairobi. Paper presented at the Regional Workshop on "Urban Policy Implications of Enhancing Food Security in African Cities", Nairobi, 27-31 May, 2002.

JACOBI, P. (1997), *Importance of vegetable production in Dar es Salaam, Tanzania*. Dar es Salaam: Urban Vegetable Promotion Project.

JACOBI, P., J. AMEND & S. KIANGO (2000), Urban agriculture in Dar es Salaam: Providing an indispensable part of the diet. In: N. Bakker *et al.*, eds., *Growing cities, growing food. Urban agriculture on the policy agenda*, pp. 257-283. Feldafing (Germany): Deutsche Stiftung für internationale Entwicklung (DSE).

JONES, S. (1999), Defining urban poverty: An overview. In S. Jones & N. Nelson, eds., *Urban poverty in Africa. From understanding to alleviation*, pp. 9-15. London: Intermediate Technology Publications.

KAAG, M., R. VAN BERKEL, J. BRONS, M. DE BRUIJN, H. VAN DIJK, L. DE HAAN, G. NOOTEBOOM & A. ZOOMERS (2004), Ways forward in livelihood research. In: D. Kalb, W. Pansters & H. Siebers, eds., *Globalization and development: Themes and concepts in current research*, pp. 49-74. Dordrecht: Kluwer Academic Publishers.

KANJI, N. (1995), Gender, poverty and economic adjustment in Harare, Zimbabwe. *Environment and Urbanization* 7(1), 37-55.

KANJI, N. (1996), Review of urbanization issues affecting children and women in the eastern and southern African region. New York: Unicef.

KARANJA, Mr. (1994, 1999), Training Commandant of Nairobi City Council Training School, personal communication.

KENYA, REPUBLIC OF (1970), *Kenya population census 1969, Vol. 1.* Nairobi: Government Printer.

KENYA, REPUBLIC OF (1981), *Kenya population census 1979, Vol. 1.* Nairobi: Government Printer.

KENYA, REPUBLIC OF (1994), *Kenya population census 1989, Vol. 1.* Nairobi: Government Printer.

KENYA, REPUBLIC OF (2000), *Economic survey 2000.* Nairobi: Government Printer.

KENYA, REPUBLIC OF (2001), *Poverty reduction strategy paper for the period 2001-2004.* Nairobi: Ministry of Finance and Planning.

KIANGO, S. & T. LIKOKO (1996), *Vegetable production on open spaces in Dar esSalaam city.* Dar es Salaam: Urban Vegetable Promotion Project.

KING'ORI, P. (forthcoming), *Food security, child nutritional status and incomes of urban farming households in Nakuru town, Kenya: A comparative study.* Nairobi: University of Nairobi, Applied Nutrition Programme, MSc thesis.

KLAVER, W. & R.K.N. MWADIME (1998), Food consumption and nutrition in the Kenya Coast. Leiden: African Studies Centre, Working Paper 31.

KULSHRETSTHA, N. (1998), Nakuru. An African town today. In: B. de Meulder et al., *Nakuru. An African town*, pp. 35-74. Leuven: City of Leuven/Catholic University.

LADO, C. (1990), Informal urban agriculture in Nairobi, Kenya: Problem or resource in development and land use policy planning. *Land Use Policy*, July 1990: 257-266.

LEE-SMITH, D., M. MANUNDU, D. LAMBA & P.K. GATHURU (1987), *Urban food production and the cooking fuel situation in urban Kenya.* Nairobi: Mazingira Institute.

LEE-SMITH, D. & P.A. MEMON (1994), Urban agriculture in Kenya. In: A.G. Egziabher *et al.*, *Cities feeding people: An examination of urban agriculture in East Africa*, pp. 67-84. Nairobi/Ottawa: IDRC.

LOURENÇO-LINDELL, I. (1996), How do the urban stay alive? Food production in a squatter settlement of Bissau. *African Urban Quarterly* 11(2-5): 163-168.

MARTIN, A., N. OUDWATER & K. MEADOWS (2000), Urban agriculture and the livelihoods of the poor in Southern Africa. Paper presented at the international symposium on "Urban agriculture and horticulture: The linkage with urban planning", Berlin, Domäne Dalem, 7-9 July 2000.

MATSHALAGA, N. (1996), "I am hungry mum!" Gender, urban food security and coping strategies. *SAFERE*, 2(1): 58-77.

MAXWELL, D. (1994), The household logic of urban farming in Kampala. In: A.G. Egziabher *et al.*, *Cities feeding people: An examination of urban agriculture in East Africa*, pp. 47-65. Nairobi/Ottawa: IDRC.

MAXWELL, D. (1995), Alternative food security strategy: A household analysis of urban agriculture in Kampala. *World Development* 23(10): 1669-1681.

MAXWELL, D. & T. FRANKENBERGER (1992), *Household food security: concepts, indicators, measurements. A technical review*. New York/Rome: UNICEF/IFAD.

MBAYE, A. & P. MOUSTIER (2000), Market-oriented urban agriculture production in Dakar. In: N. Bakker *et al.*, eds., *Growing cities, growing food. Urban agriculture on the policy agenda*, pp. 235-256. Feldafing (Germany): Deutsche Stiftung für internationale Entwicklung (DSE).

MBIBA, B. (1995), *Urban agriculture in Zimbabwe: Implications for urban management and poverty*. Aldershot: Avebury.

MBIBA, B. (2000), Urban agriculture in Harare: Between suspicion and repression. In: N. Bakker *et al.*, eds., *Growing cities, growing food. Urban agriculture on the policy agenda*, pp. 285-301. Feldafing (Germany): Deutsche Stiftung für internationale Entwicklung (DSE).

MCN (1999), *Strategic Nakuru Structure Plan. Action plan for sustainable urban development of Nakuru town and its environs*. Nakuru: Municipal Council of Nakuru (Final draft).

MEIKLE, S. (2002), The urban context and poor people. In: C. Rakodi, with T. Lloyd-Jones, eds., *Urban livelihoods: A people-centred approach to reducing poverty*, pp. 37-51. London: Earthscan Publications.

MEME, M.M. (1996), *Nutrient intake and nutritional status of children in a school with and without a feeding programme in Nyambene District, Kenya*. Nairobi: University of Nairobi, Applied Nutrition Programme, MSc thesis.

MEMON, P.A. & D. LEE-SMITH (1993), Urban agriculture in Kenya. *Canadian Journal of African Studies* 27(1): 25-42.

MIANDA, G. (1996), Women and garden produce of Kinshasa: The difficult quest for autonomy. In: Parvin Gharayshi & Claire Belanger, eds., *Women, work and gender relations in developing countries*. Westport, Conn.: Greenwood Press.

MIRERI, C. (2002), The implication of planning & management on urban agriculture in Kenya: Opportunities and challenges. Paper presented at the Regional Workshop on "Urban Policy Implications of Enhancing Food Security in African Cities", Nairobi, 27-31 May, 2002.

MITLIN, D. (2003), Addressing urban poverty through strengthening assets. *Habitat International* 27(2003): 393-406.

MKAWALE, S. (2000), Nakuru rivers writhing under massive sludge. East African Standard, September 25, 2000, "The Big Issue", p. 3.

MLOZI, M.R.S. (1996), Urban agriculture in Dar es Salaam: Its contribution to solving the economic crisis and the damage to its environment. *Development Southern Africa* 13(1): 47-65.

MLOZI, M.R.S. (1997), Urban agriculture: Ethnicity, cattle raising and some environmental implications in the City of Dar es Salaam, Tanzania. *African Studies Review* 40(3): 1-28.

MLOZI, M.R.S. (1998), Urban vegetable production in low-density areas of Dar es Salaam. Dar es Salaam: Urban Vegetable Promotion Project.

MLOZI, M.R.S. (1999), Urban agriculture and its damaging effects on the urban environment: The case of Tanzania. In P.G. Forster & S. Maghimbi, eds., *Agrarian economy, state and society in contemporary Tanzania*, pp. 184-199. Aldershot: Ashgate.

MLOZI, M.R.S., I.J. LUPANGA & Z.S.K. MVENA (1992), Urban agriculture as a survival strategy in Tanzania. In: J. Baker & P.O. Pedersen, eds., *The rural-urban interface in Africa: Expansion and adaptation*, pp. 284-294. Uppsala: The Scandinavian Institute of African Studies.

MOSHA, A.C. (1991), Urban farming practices in Tanzania. *Review of Rural and Urban Planning in Southern and Eastern Africa* 1: 83-92.

MOUGEOT, L.J.A. (1993), Urban food self-reliance: significance and prospects'. *IDRC Reports* 21(3): 2-5.

MOUGEOT, L.J.A. (1994a), African city farming from a world perspective. In: A.G. Egziabher et al., *Cities feeding people. An examination of urban agriculture in East Africa*, pp. 1-24. Ottawa: International Development Research Centre.

MOUGEOT, L.J.A. (1994b), *Urban food production: Evolution, official support and significance*. Ottawa: IDRC, Cities Feeding People Series, Report 8.

MOUGEOT, L.J.A. (2005), Introduction. In L.J.A. Mougeot, ed., *Agropolis. The social, political and environmental dimensions of urban agriculture*, pp. 1-29. London & Ottawa: Earthscan & International Development Research Centre (IDRC).

MUBVAMI, T., S. MUSHAMBA & R. VAN VEENHUIZEN (2003), Availability, access and usability of land for urban agriculture. *Urban Agriculture Magazine* 11: 1-3.

MUGAMBI, A.M. (2002), Urban agriculture in Kenya: Technical and policy issues. Paper presented at the Regional Workshop on "Urban Policy Implications of Enhancing Food Security in African Cities", Nairobi, 27-31 May, 2002.

MUKUI, J.T. (2002), Urban and peri-urban agriculture and rural-to-urban food flows: Case study of Nairobi. Paper presented at the Regional Workshop on "Urban Policy Implications of Enhancing Food Security in African Cities", Nairobi, 27-31 May, 2002.

MUNARI, Mr. (1994), Nairobi City Council Public Health Prosecution Officer, personal communication.

MUREITHI, D. (2000), manager, Nakuru Prison Farm, personal communication, 20-09-2000.

MUSHAMBA, S. (2002), Policy and planning options for urban agriculture relevant for the Nakuru Urban Agriculture research Project. Keynote address presented at the Nakuru Urban Agriculture research Project workshop, 27-28 November 2002.

MUSTER, G. (1997), *Environmental problems of urban agriculture: A case study of Dar es Salaam, Tanzania*. Dar es Salaam: Urban Vegetable Promotion Project.

MWAGO, G. (2000), SENVINET, Nakuru, personal communication, 13-09-2000.

MWANGI, A.M. (1995), *The role of urban agriculture in food security in low-income areas in Nairobi, Kenya*. Leiden: African Studies Centre.

MWANGI, A.M. & D. FOEKEN (1996), Urban agriculture, food security and nutrition in low-income areas in Nairobi, Kenya. *African Urban Quarterly* 11(2-3): 170-179.

MWANGI, I.K. (2002), Using urban planning instrument in improving urban food security. Paper presented at the Regional Workshop on "Urban Policy Implications of Enhancing Food Security in African Cities", Nairobi, 27-31 May, 2002.

MWANGI, S.W. (2001), *Local Agenda 21 experiences in Nakuru, Kenya: Processes, issues and lessons*. London: International Institute for Environment and Development (IIED), Urban Environmental Action Plans & Local Agenda 21 Series, Working Paper 10.

NELSON, T. (1996), Closing the nutrient loop. *World Watch*, November/December 1996, pp. 10-17.

NKONYA, D.M. (1997), *Effectiveness of urban agriculture/livestock extension agents in disseminating information for reducing environmental damage that urban agriculture causes*. Morogoro: Sokoine University of Agriculture, Department of Agricultural Education and Extension, unpublished BSc. special project.

NUGENT, R. (2000), The impact of urban agriculture on the household and local economies. In: N. Bakker *et al.*, eds., *Growing cities, growing food. Urban agriculture on the policy agenda*, pp. 67-98. Feldafing (Germany): Deutsche Stiftung für internationale Entwicklung (DSE).

NUWAGABA, A. (1996), Urbanisation and environmental crisis in a Ugandan city: implications for environment management and sustainable development. *Eastern Africa Social Science Research Review*, 12(1): 15-34.

NYANDWARO, E.O. (forthcoming), *Environmental impacts of urban farming: A case study of Nakuru town, Kenya*. Nairobi: Kenyatta University, MSc thesis.

OBOSU-MENSAH, K. (1999), *Food production in urban areas. A study of urban agriculture in Accra, Ghana*. Aldershot: Ashgate.

OBUDHO, R.A. & D. FOEKEN (1999), *Urban agriculture in Africa. A bibliographical survey*. Leiden/Nairobi: African Studies Centre/Centre for Urban Research, ASC Research Report 58.

ODERA, C. (forthcoming), *The implications of urban school farming on food security and nutrition: A study in Nakuru town*. Nairobi: University of Nairobi, Applied Nutrition Programme, MSc thesis.

OWUOR, S.O. (2002), Enhancing food security in African cities: Rural farming by urban households – The case of Nakuru town, Kenya. Paper presented at the "Workshop on urban policy implications of enhancing food security in African cities", Nairobi, 27-31 May 2002.

OWUOR, S.O. (2003), *Rural livelihood sources for urban households. A study of Nakuru town, Kenya*. Leiden: African Studies Centre, Working Paper 51.

OWUOR, S.O. (2006), *Bridging the urban-rural divide: Multi-spatial livelihoods in Nakuru town, Kenya*. Leiden: African Studies Centre.

OWUOR, S.O. & D. FOEKEN (2006), Surviving in the neighbourhoods of Nakuru town, Kenya. In P. Konings & D. Foeken, eds., *Crisis and creativity. Exploring the wealth of the African neighbourhood*. Leiden: Brill.

PAYNE, G. (2002), Tenure and shelter in urban livelihoods. In: C. Rakodi, with T. Lloyd-Jones, eds., *Urban livelihoods: A people-centred approach to reducing poverty*, pp. 151-164. London: Earthscan Publications.

PIETERS, J., J. DE MOEL, O. VAN STEENBERGEN & W. VAN DER HOEVEN (1977), Effects of school feeding on growth of children in Kirinyaga District, Kenya. *East African Medical Journal* 54: 622-630.

PLATT, B.S. (1962), *Tables of representative values of foods commonly eaten in tropical countries*. London: Her Majesty's Stationary Office, Medical Research Council Special Report Series no. 302.

POTTS, D. & C. MUTAMBIRWA (1990), Rural-urban linkages in contemporary Harare: why migrants need their land. *Journal of Southern African Studies*, 16(4): 677-698.

POTUTAN. G., E.M. IMAM, R.J. HOLMER & W.H. SCHNITZLER (1999), School gardens of Cagayan de Oro contributing to food security. Initial observations (unpublished).

QUON, S. (1999), *Planning for urban agriculture: A review of tools and strategies for urban planners*. Ottawa: IDRC, Cities Feeding People Report Series, Report 28.

RAKODI, C. (1988), Urban agriculture: Research questions and the Zambian experience. *Journal of Modern African Studies* 26(3): 495-515.

RAKODI, C. (1995), The household strategies of the urban poor: Coping with poverty and recession in Gweru, Zimbabwe. *Habitat International* 19(4): 447-471.

RAKODI, C. (2002a), A livelihoods approach – Conceptual issues and definitions. In C. Rakodi, with T. Lloyd-Jones, eds, *Urban livelihoods: A people-centred approach to reducing poverty*, pp. 3-22. London: Earthscan Publications.

RAKODI, C. (2002b), Conclusions. In C. Rakodi, with T. Lloyd-Jones, eds, *Urban livelihoods: A people-centred approach to reducing poverty*, pp. 288-298. London: Earthscan Publications.

ROGERSON, C.M. (1992), Feeding Africa's cities: the role and potential for urban agriculture. *Africa Insight*, 22(4): 229-234.

ROGERSON, C.M. (1994), Urban agriculture in South Africa: Scope, issues and potential. *GeoJournal* 30: 21-28.

SANYAL, B. (1985), Urban agriculture: Who cultivates and why? A case study of Lusaka, Zambia. *Food and Nutrition Bulletin* 7(3): 15-24.

SAWIO, C. (1993), *Feeding the urban masses? Towards an understanding of the dynamics of urban agriculture and land use change in Dar es Salaam*. Worcester: Clark University, unpublished PhD thesis.

SAWIO, C. (1994), Who are the farmers of Dar es Salaam? In: Axumite G. Egziabher *et al.*, *Cities feeding people: An examination of urban agriculture in East Africa*, pp. 25-46. Nairobi/Ottawa: IDRC.

SAWIO, C. (1996), *Urban agriculture in Dar es Salaam: Environmental planning and management process*. Interim technical report to IDRC, Dar es Salaam, Tanzania (unpublished).

SCHILTER, C. (1991), L'agriculture urbaine: Une activité creative d'emplois en économie de survie. Le cas de Lomé. *Cahiers des Sciences Humaines* 27(2): 159-168.

SEHMI, J.K. (1993), *National food composition tables and the planning of satisfactory diets in Kenya*. Nairobi: Government Printer.

SEMWANGA, M.A. (2002), The magnitude of urban food poverty in Uganda. The case of Kampala. Paper presented at the Regional Workshop on "Urban Policy Implications of Enhancing Food Security in African Cities", Nairobi, 27-31 May, 2002.

SEN, A. (1981), *Poverty and famines. An essay on entitlement and deprivation*. Oxford: Oxford University Press.

SHAURI, R. (1988), *Rus in Urbe: A study of urban agriculture in Dar es Salaam*. Dar es Salaam: Ardhi Institute, unpublished diploma project.

SHELDON, K. (1991), *Farming in the city: Urban women in agricultural work in Mozambique*. Los Angeles: University of California.

SMIT, J., A. RATTA & J. NASR (1996), *Urban agriculture: Food, jobs and sustainable cities*. New York: United Nations Development Programme (UNDP).

STEVENSON, C., P. XAVERY & A. WENDELINE (1996), *Market production of fruits and vegetables in the peri-urban area of Dar es Salaam*. Dar es Salaam: Urban Vegetable Promotion Project.

STREIFFELER, F. (1994), Urban agriculture in Africa: The recent situation in its most important aspects. *Africa Hilbt Sich Selbst*, 4: 241-255.

TELLEGEN, N. (1997), *Rural enterprises in Malawi: Necessity or opportunity?* Aldershot: Ashgate.

TINKER, I. (1994), Urban agriculture is already feeding cities. In: A.G. Egziabher et al., *Cities feeding people. An examination of urban agriculture in East Africa*, pp. vii-xiv. Ottawa: International Development Research Centre.

TRICAUD, P.M. (1987), *Urban agriculture in Ibadan and Freetown*. Tokyo: United Nations University, Food Energy Nexus Programme.

TUTS, R. (1998), Localizing Agenda 21 in small cities in Kenya, Morocco and Vietnam. *Environment and Urbanization*, 10(2), 175-189.

UNCHS (1996), *An urbanizing world: global report on human settlements, 1996*. London/Nairobi: Oxford University Press/United Nations Centre for Human Settlements (Habitat).

VAN DE HOEK, W. (2002), The use of untreated wastewater in agriculture: Strategies for health risk management. Discussion paper for the RUAF-IWMI e-mail conference on "Use of untreated urban wastewater in agriculture in low income countries".

VENNETIER, P. (1961), La vie agricole urbaine à Pointe Noire (Congo). *Cahiers d'Outre-Mer* 14(55): 60-84.

VERSLEIJEN, N. (2002), *Sukuma! A social analysis of urban agriculture: Case studies from Nakuru Town, Kenya.* Wageningen: Wageningen University and Research Center, Department of Rural Development Sociology, MSc thesis.

VILLIEN, F. (1988), L'agriculture dans la ville: L'exemple de Bangui. *Cahiers d'Outre-Mer* 41(163): 163-180.

WALLMAN, S. (1996), *Kampala women getting by: Well-being in the time of AIDS*. London: James Currey.

YACHKASCHI, J. (1997), *Urban and peri-urban production, marketing system and consumption of fruit and vegetables in selected cities in Tanzania.* Dar es Salaam: Urban Vegetable Promotion Project.

Index

AFRIKA-STUDIECENTRUM SERIES

ISSN 1570–9310

1. Konings, P. and Nyamnjoh, F.B. *Negotiating an Anglophone Identity.* A Study of the Politics of Recognition and Representation in Cameroon. 2003. ISBN 90 04 13295 3
2. Wels, H. *Private Wildlife Conservation in Zimbabwe.* Joint Ventures and Reciprocity. 2003. ISBN 90 04 13697 5.
3. Spierenburg, M.J. *Strangers, Spirits and Land Reforms.* Conflicts about Land in Dande, Northern Zimbabwe. 2004. ISBN 90 04 13657 5.
4. Müller, T.R. *The Making of Elite Women.* Revolution and Nation Building in Eritrea. 2005. ISBN 90 04 14287 8.
5. Ntsebeza, L. *Democracy Compromised.* Chiefs and the politics of the land in South Africa. 2005. ISBN 90 04 14482 X.
6. Evers, S., Spierenburg, M. and Wels, H. *Competing Jurisdictions.* Settling land claims in Africa. 2005. ISBN 90 04 14780 2.
7. Foeken, D. "*To Subsidise My Income*". Urban Farming in an East-African Town. 2006. ISBN-13: 978-90-04-15202-1, ISBN-10: 90-04-15202-4.